ALEXANDER SOIFER

University of Colorado
at Colorado Springs

MATHEMATICS AS PROBLEM SOLVING

With 67 Illustrations, Tables, Diagrams

Center for Excellence in Mathematical Education
Colorado Springs, 1987

Center for Excellence in Mathematical Education,
885 Red Mesa Drive
Colorado Springs, Colorado 80906, U.S.A.

Author: Alexander Soifer, Professor of Mathematics,
Department of Mathematics, University of Colorado
Colorado Springs, Colorado 80933, U.S.A.

Cover: Yuri Soifer
Illustrations: Alexander Soifer

First Edition

Printed in the United States of America

AMS Subject Classification (1981):00A07
Library of Congress Catalog Card Number 86-072215
ISBN 0-940263-00-9

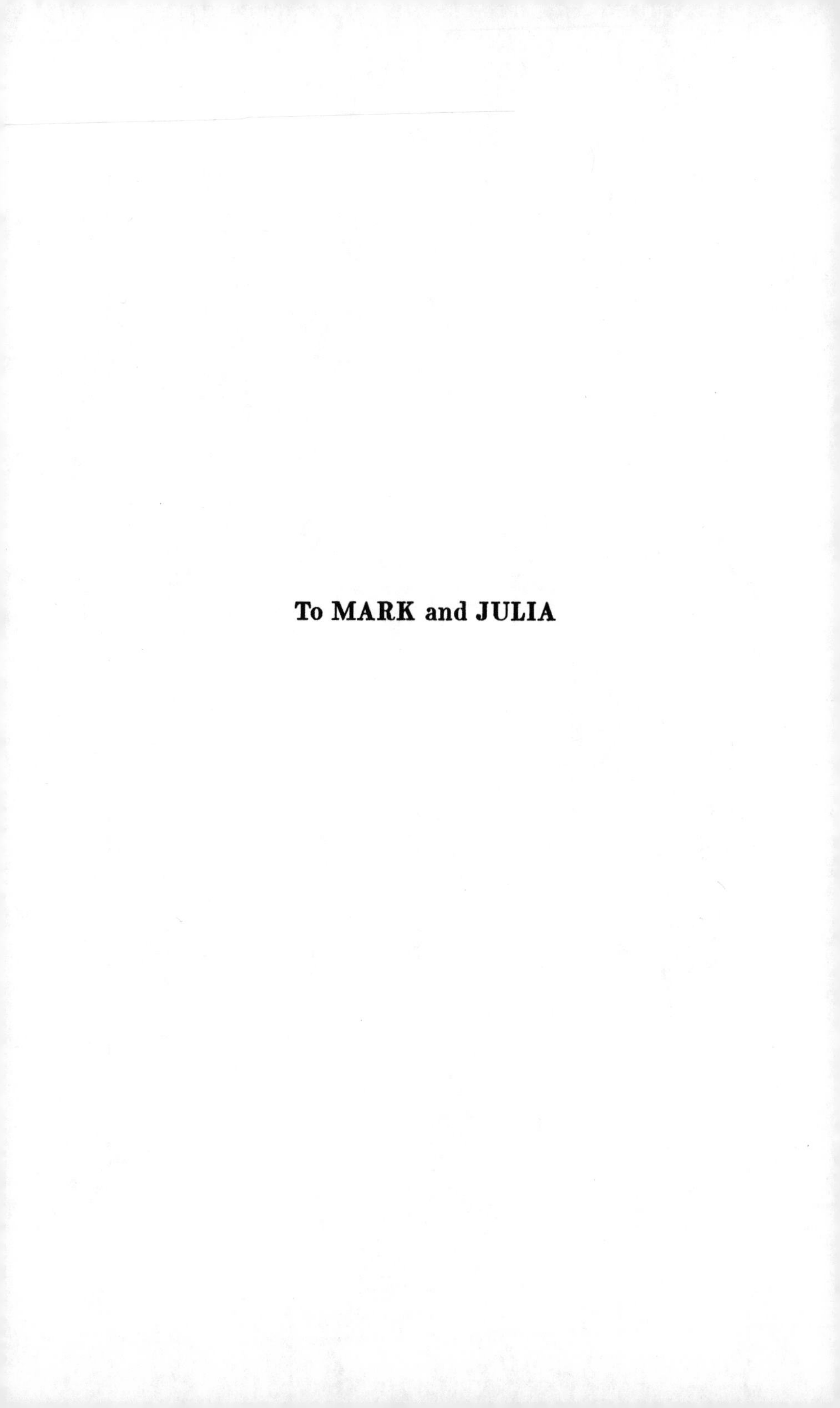

To MARK and JULIA

Preface

Remember but him, who being demanded, to what purpose he toiled so much about an Art, which could by no means come to the knowledge of many. Few are enough for me; one will suffice, yea, less than one will content me, answered he. He said true: you and another are a sufficient theatre one for another; or you to your selfe alone!!

Michel de Montaigne.
Of Solitarinesse. The Essayes

I was fortunate to grow up in a problem solving atmosphere of Moscow with its mathematical clubs, schools, and olympiads. The material for this book stems from my participation in numerous mathematical competitions of all levels, from school to National, as a competitor, an organizer, a judge, and a problem writer; but most importantly, from the mathematical folklore I grew up on.

This book contains about 200 problems, over one-third of which is discussed in detail, sometimes even with two or more solutions. When I started, I thought that beauty, challenge, and paradoxy of the results and solutions alone would determine my choices. During my work, however, one more factor powerfully forced itself into account: interplay of selected problems.

This book is written for high school and college students, teachers, and everyone else desiring to experience the mystery and beauty of mathematics. It can be and has been used as a text for an undergraduate or graduate course or workshop on problem solving.

Auguste Renoir once said that just as some people all their lives read one book (the Bible, for example), so can he paint all his life one painting. I cannot agree with him more. This is the book I am going to write all my life. That is why I welcome so much your comments, corrections, ideas, alternative solutions, suggestions to include other methods, or to cover other areas of mathematics. Do send me your ideas and solutions: best of them as well as the names of their authors will be included in the future revised editions of this book. I hope though that this book will never reach the intimidating size of a calculus text.

One can fairly make an argument that this book is raw, unpolished. Perhaps it is not all that bad: sketches of Modigliani give me, for one, so much more than sweated out oils of Old Masters. Maybe a problem solving book ought to be a sketch book!

To assign a true authorship to these problems is as difficult as to folklore tales. A few references that I have given, indicate my source rather than a positive reference to the first mentioning of a problem. Even problems that I created and published myself might prove to have existed before I was born!

I thank Valarie Barnes for bravely agreeing to type this manuscript: it was her first encounter of the mathematical kind. I thank my student Richard Jessop for producing such a masterpiece of typesetting art.

I am grateful to my parents Yuri and Rebbeca for introducing me to the world of arts, and to my children Mark and Julia for inspiration. My special thanks go to the first judges of this manuscript, my students in Colorado Springs and Southampton for their enthusiasm, ideas, and support.

Colorado Springs A. SOIFER
November 1986

Table of Contents

Chapter 5 Combinatorial Problems

Literature

Chapter 1
Language
and
A Few Celebrated Ideas

1.1 Streetcar Stories

I would like to start our discussion with the following story.

Streetcar Story I

You enter a streetcar with six other passengers on the first stop of its route. On the second stop four people come in and two get off. On the third stop seven people come in and five get off. On the fourth stop eight people come in and three get off. On the fifth stop thirteen people come in and eight get off.

How old is the driver?

Did you start counting passengers in the streetcar? If you did, here is your first lesson:

Do not start solving a problem before you read it!

Sounds obvious? Perhaps you are right. But you should not underestimate the importance of it. I, for one, underestimated some obvious things in life, and had to learn the hard way lessons like "do not read while you drive"!

The story above does not give us any information relevant to the age of the driver. However, relevance of information is not at all always obvious.

Streetcar Story II

The reunion of two friends in a streetcar sounded like this:

— How are you, thank you I am fine.

— You just got married when we met last. Any children?

— I have three kids!

— Wow! How old are they?

— Well, if you multiply their ages, you would get 36; but if you add them up, you'd get the number of passengers in this streetcar.

— Gotcha, but you did not tell me enough to figure out their ages.

— My oldest kid is a great sportsman.

— Aha! Now I know their ages!

Find the number of passengers in the streetcar and the ages of the children.

Can the statement "my oldest kid is a great sportsman" have any relevance? It can. In fact, it does! Moreover, the fact that without this statement the second friend cannot figure out the ages of the children carries a valuable information too!

Let us just take a look at the following table:

Decompositions of 36 into 3 factors x, y, z	The sum $x + y + z$
$1 \cdot 1 \cdot 36$	38
$1 \cdot 2 \cdot 18$	21
$1 \cdot 3 \cdot 12$	16
$1 \cdot 4 \cdot 9$	14
$1 \cdot 6 \cdot 6$	13
$2 \cdot 2 \cdot 9$	13
$2 \cdot 3 \cdot 6$	11
$3 \cdot 3 \cdot 4$	10

The fact that the second friend was unable to figure out the ages x, y, z of the children when he knew their sum $x + y + z$ implies that there must be at least two solutions for the given sum $x + y + z$ of ages! The table shows that only 13 appears twice in the column "$x + y + z$," therefore, $x + y + z = 13$, and we know the number of passengers!

We can also see the relevance of the oldest kid being a great sportsman: it rules out 1,6,6 and leaves 2,2,9!

1.2 Language

As any other science, mathematics is formulated in an ordinary language, English in the United States. It is very essential to use the language correctly as well as to correctly interpret sentences. I have no intention to discuss in these lectures formal mathematical languages. I would like, however, to briefly talk about constructing complex sentences, and to define the meaning of the words "not", "and,", "or," "imply," "if and only if," etc.

We will deal only with statements which are clearly true or false in a given context.

Here are a few examples of such statements:

(1) Chicago is the capital of the United States.
(2) One yard is equal to three feet.
(3) Any sports car is red.
(4) Any Ferrari is red.

As you can see, the first and third statements are false and the second statement is true. It took me a visit to my friend Bob Penkhus, a car dealer, to find out that the fourth statement is false.

The truth or falsity of a composite statement is completely determined by the truth or falsity of its components.

1. Negation

Given a statement A. The negation of A, denoted by $\neg A$ and read "not A", is a new statement, which is understood to assert that "A is false."

Let 1 stand for truth, and 0 stand for false, then the following table defines the values of $\neg A$:

A	$\neg A$
1	0
0	1

i.e., $\neg A$ is false when A is true, and $\neg A$ is true when A is false.

2. Conjunction

Given statements A and B. The conjunction of A and B, denoted $A \wedge B$ and read "A and B", is a new statement, which is understood to assert that "A is true and B is true". The following truth table defines $A \wedge B$:

A	B	$A \wedge B$
1	1	1
1	0	0
0	1	0
0	0	0

3. Disjunction

Given statements A and B. The disjunction of A and B, denoted $A \vee B$, and read "A or B", is a new statement, which is understood to assert that "at least one of the statements A, B is true". The following truth table defines $A \vee B$:

A	B	$A \vee B$
1	1	1
1	0	1
0	1	1
0	0	0

4. Implication

Given statements A and B. The implication $A \Rightarrow B$, to be read "A implies B", is a new statement which is understood to assert that "if A is true, than B is true". It is defined by the following truth table:

A	B	$A \Rightarrow B$
1	1	1
1	0	0
0	1	1
0	0	1

Please note that the meaning of the word implication in mathematics is quite different from the common language use of this word: $A \Rightarrow B$ is false only if A is true and B is false.

5. Equivalence

Given statements A and B. The equivalence $A \Leftrightarrow B$, to be read "A equivalent B", is an abbreviation for the following statement:

$$(A \Rightarrow B) \wedge (B \Rightarrow A)$$

In order to uniquely interpret a composite statement, we would need to use sometimes lots of parentheses. This can make a statement quite difficult to read or visually evaluate. We can resolve this problem in exactly the same way as we do it in arithmetic: by establishing the order of operations in a parenthesis-free composite statement:

We apply	$\neg$	first
	$\wedge$	second
	$\vee$	third
	$\Rightarrow$	fourth
	$\Leftrightarrow$	fifth

Finally, a composite statement which is true regardless of the truth or falsity of its components, is called *tautology.*

Problems

Prove the following tautologies:

1.2.1 $A \Rightarrow A$

1.2.2 $A \Rightarrow A \vee B$

1.2.3 $A \wedge B \Rightarrow A$

1.2.4 $\neg\neg A \Leftrightarrow A$

1.2.5 $\neg(A \vee B) \Leftrightarrow \neg A \wedge \neg B$ (De Morgan's Law)

1.2.6 $\neg(A \wedge B) \Leftrightarrow \neg A \vee \neg B$ (De Morgan's Law)

1.2.7 $(A \Rightarrow B) \wedge (B \Rightarrow C) \Rightarrow (A \Rightarrow C)$

1.2.8 $(\neg B \Rightarrow \neg A) \Leftrightarrow (A \Rightarrow B)$

1.2.9 $(A \wedge \neg B \Rightarrow \neg A) \Rightarrow (A \Rightarrow B)$

1.2.10 $(A \wedge \neg B \Rightarrow \mathcal{F}) \Rightarrow (A \Rightarrow B)$, ($\mathcal{F}$ denotes a false statement)

From now on we will use symbols:

$\wedge$	for "and"
$\vee$	for "or"
$\Rightarrow$	for "implies"
$\Leftrightarrow$	for "if and only if"
$\neg$	for "not"
$\exists$	for "there exists"
$\forall$	for "for every".

If $A \Rightarrow B$ is true, we would say that A is a sufficient condition for B; at the same time we would say, that B is a necessary condition for A. If $A \Leftrightarrow B$ is true, than B is said to be a necessary and sufficient condition for A. Please understand and remember that a statement *converse* to $A \Rightarrow B$ is $B \Rightarrow A$. A statement *opposite* to $A \Rightarrow B$ is $\neg(A \Rightarrow B)$.

1.3 Arguing By Contradiction

Problems 1.2.9 and 1.2.10 presented the following tautologies:

$$(A \wedge \neg B \Rightarrow \neg A) \Rightarrow (A \Rightarrow B)$$
$$(A \wedge \neg B \Rightarrow \mathcal{F}) \Rightarrow (A \Rightarrow B).$$

These two tautologies justify a celebrated method of mathematical proofs: arguing by contradiction.

Let us say we are given that A is true, and we are asked to prove that B is true. We assume that B is not true (i.e., $\neg B$ is true), and then start with A and $\neg B$ and continue deductions until we arrive at a contradiction to what is given (i.e. at $\neg A$), or to what is known to be true.

1.3.1 Prove that the sum of a rational number and an irrational number is again an irrational number.

Proof

Let r be a rational number (i.e., $r = m/n$ for some integers m, n with $n \neq 0$), and i be an irrational number (i.e., i cannot be presented in the form s/t, where s, t are integers and $t \neq 0$).

Assume that the sum $r + i$ is a rational number, say r_1, then if $r_1 = p/q$ for integers $p, q; q \neq 0$, we get:

$$r + i = r_1$$
$$i = r_1 - r$$
$$i = \frac{p}{q} - \frac{m}{n}$$
$$i = \frac{np - mq}{nq}$$

i.e. i is a rational number, which contradicts to the given fact that i is an irrational number. Therefore, $r + i$ is irrational.

■

1.3.2 (*Pigeonhole Principle*) If $kn + 1$ pigeons (k, n are positive integers) are placed in n pigeonholes, then at least one of the holes contains at least $k + 1$ pigeons.

Proof

Assume that there are no holes which contain at least $k + 1$ pigeons. Then

$$\begin{array}{rll} & \text{the 1st hole contains} & \leq k \text{ pigeons} \\ & \text{the 2nd hole contains} & \leq k \text{ pigeons} \\ & \vdots & \\ + & \text{the } n\text{th hole contains} & \leq k \text{ pigeons} \\ \hline & \text{total number of pigeons} & \leq k \times n \end{array}$$

This contradicts to the given fact that there are $kn + 1$ pigeons. Therefore, there is a hole which contains at least $k + 1$ pigeons.

∎

Problems

1.3.3 Prove that the product of a non-zero rational number and an irrational number is again an irrational number.

1.3.4 Given a prime p and a positive integer n. Prove that if n^2 is divisible by p, then n is divisible by p. (Hint: The Fundamental Theorem of Arithmetic states that any positive integer greater than 1 can be decomposed into the product of prime numbers, and this decomposition is unique up to the order of factors.)

1.3.5 Prove that $\sqrt{6}$ is an irrational number.

1.3.6 A known theorem states that any point $\mathcal{C}$ of the perpendicular bisector of a segment $\overline{\mathcal{AB}}$ is equidistant from $\mathcal{A}$ and $\mathcal{B}$. Prove the converse.

1.3.7 A known theorem states that if a convex quadrilateral is inscribed in a circle, then the sums of its opposite angles are equal. Prove the converse.

1.3.8 A known theorem states that $m = \frac{a+b}{2}$, where m is the length of the median and a, b are the lengths of the parallel bases of a trapezoid. Prove the converse.

1.4 Pigeonhole Principle

In Section 1.3. we already proved the Pigeonhole Principle, also known as the Dirichlet Principle (after its inventor, a famous mathematician Peter Gustav Dirichlet, 1805–1859). This simple principle does wonders. It is amazing how easy it is to understand this idea, and how difficult sometimes it is to discover that this idea can be applied! That is why in April 1986, I called my lecture for participants of the Colorado Springs Mathematical Olympiad "Invisible Pigeonhole Principle".

When you look at the problems that are solved by the Pigeonhole Principle, many of them appear to have nothing in common.

1.4.1 New York City has 7,500,000 residents. Maximal number of hairs that can grow on a human head is 500,000. Prove that there are at least 15 residents in New York City with the same number of hairs.

Solution

Let us set up 500,001 pigeonholes labeled by integers 0 to 500,000, and put residents of New York into the holes labeled by the number of hairs on their heads.

Due to $7,500,000 > 14 \times 500,001 + 1$ we conclude by the Pigeonhole Principle that there is a pigeonhole with at least $14 + 1$ "pigeons", i.e., there are at least 15 residents of New York with the same number of hairs.

■

1.4.2 (Third Annual Colorado Springs Mathematical Olympiad, 1986). Santa Claus and his elves paint the plane in two colors, red and green. Prove that there exist two points of the same color exactly one mile apart.

Solution

Consider an equilateral triangle with sides equal to one mile on the given plane. Since its three vertices (pigeons) are painted in two colors (pigeonholes), we can choose two vertices painted in the same color (at least two pigeons in a hole).

■

1.4.3 (Third Annual Colorado Springs Mathematical Olympiad, 1986) Given n integers, prove that either one of them is a multiple of n, or some of them add up to a multiple of n.

Solution

Denote given integers by $a_1, a_2, \ldots, a_n$. Define:

$$\begin{aligned} S_1 &= a_1 \\ S_2 &= a_1 + a_2 \\ &\vdots \\ S_n &= a_1 + a_2 + \cdots + a_n \end{aligned}$$

If one of the numbers $S_1, S_2, \ldots, S_n$ is a multiple of n, we are done. Assume now that none of the numbers $S_1, S_2, \ldots, S_n$ is a multiple of n. Then all possible remainders upon the division of these numbers by n are $1, 2, \ldots, n-1$, i.e., we get more numbers (namely n, they are our "pigeons") than possible remainders ($n-1$ possible remainders, they are our "pigeonholes"), therefore, among the numbers $S_1, S_2, \ldots, S_n$ there exist two numbers, say S_k and S_{k+t} which give the same remainders upon the division by n.

We are done, because

(1) $S_{k+t} - S_k$ is a multiple of n;
(2) $S_{k+t} - S_k = a_{k+1} + a_{k+2} + \cdots + a_{k+t}$;

in other words, we found some of the given numbers, namely $a_{k+1}, a_{k+2}, \ldots, a_{k+t}$, whose sum is a multiple of n.

■

1.4.4 Given a real number r. Prove that among its first ninety-nine multiples $r, 2r, \ldots, 99r$ there is at least one that differs from an integer by not more than $1/100$.

Solution

Let us roll the number line on the roller with circumference equal to 1 (figure 1.1)

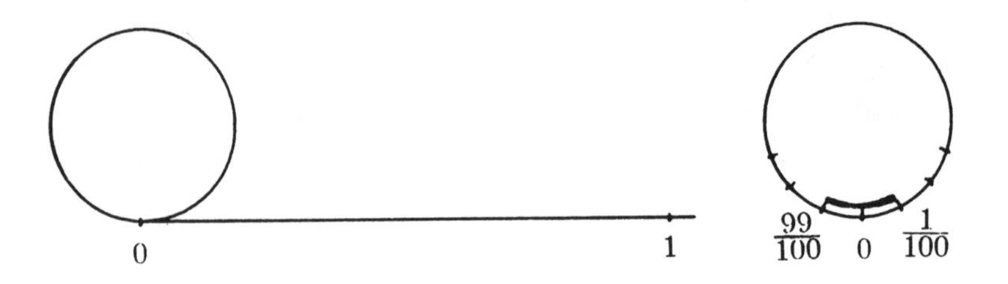

Figure 1.1 Figure 1.2

All integers will coinside on the roller with zero. Now we divide the circumference into 100 arcs of equal length (Figure 1.2).

If at least one multiple $kr, 1 \leq k \leq 99$ lies on one of the arcs $[99/100, 0]$ or $[0, 1/100]$, then we are done: kr differs from an integer by not more than $1/100$.

Assume now that none of the multiples kr, $r = 1, 2, \ldots, 99$ lies on the two arcs above. We have 99 pigeons (numbers $r, 2r, \ldots, 99r$) in $100 - 2 = 98$ pigeonholes (the remaining 98 arcs), therefore by Pigeonhole Principle at least two of the multiples, say kr and tr, $(k > t)$, lie on the same arc of the length $1/100$. All there is left to notice is:

(a) $kr - tr = (k - t)r$ is one of the given 99 multiples;

(b) $kr - tr$ lies on one of the arcs $[99/100, 0]$ or $[0, 1/100]$; which contradicts to our assumption.

■

1.4.5 (A. Soifer and S.G. Slobodnik, 1973). Forty-one rooks are placed on a 10×10 chessboard. Prove that you can choose five of them which are not attacking each other. (We say that one rook "attacks" another if they are in the same row or column of the chessboard.)

Solution

(a) Since 41 rooks (pigeons) are placed on ten rows of the board (pigeonholes), and $41 = 4 \times 10 + 1$, there exists a row A with at least five rooks on it.

If we remove row A, we will have nine rows (pigeonholes) left with at least $41 - 10 = 31$ rooks on them. Since $31 > 3 \times 9 + 1$, there is a row B among the nine rows with at least four rooks on it.

Now we remove rows A and B. We are left with eight rows (pigeonholes) and at least $41 - 2 \times 10 = 21$ rooks on them. Since $21 > 2 \times 8 + 1$, there is a row C among the eight rows with at least three rooks on it.

Continuing this reasoning, we get row D with at least two rooks on it, and row E with at least one rook on it.

(b) Now we are ready to select the required five rooks.

First we pick any rook R_1 from row E (at least one exists there, remember!).

Next we pick a rook R_2 from row D, which is not in the same column as R_1. This can be done too, because at least two rooks exist in row D.

Next, of course, we pick a rook R_3 from row C, which is not in the same row as R_1 and R_2. Conveniently, it can be done since row C contains at least three rooks.

Continuing this construction, we end up with the five rooks R_1, R_2, ..., R_5 such that they are from different rows (one per row, out of the selected rows, A, B, C, D, E) and from different columns, therefore, they do not attack each other.

■

Problems

1.4.6 A three-dimensional space is painted in three colors. Prove that there are two points one mile apart painted in the same color.

1.4.7 Given a square of the size 1×1 and five points inside it. Prove that among the given points there are two with the distance not exceeding $\frac{1}{2}\sqrt{2}$ between them.

1.4.8 A number of people (more than one) came to a party. Prove that at least two of them shook equal numbers of hands during the party.

1.4.9 (I.M. Gelfand) Little grooves of the same width are dug across a long (very long!) straight road. The distance between the centers of any two consecutive grooves is $\sqrt{2}$. Prove that no matter how narrow the grooves are, a man walking along the road with a step equal to 1, sooner or later will step into a groove.

1.4.10 (First Annual Southampton Mathematical Olympiad, 1986). Grandmaster Lev Alburt plays at least one game of chess a day to keep in shape and not more than 10 games a week not to get too tired. Prove that: if he plays long enough there will be a series of consecutive days during which he will play exactly 21 games.

1.4.11 (First Annual Southampton Mathematical Olympiad, 1986). An organization consisting of n members ($n > 5$) has $n + 1$ three-member committees, no two of which have identical membership. Prove that there are two committees which have exactly one member in common.

1.4.12 (A. Soifer and S. Slobodnik, 1973). Given 51 distinct two-digit numbers. Prove that you can choose six numbers out of them, such that any two of the six numbers have distinct digits of units, and distinct digits of tens.

1.4.13 (A. Soifer and S. Slobodnik, 1973) Given $r \cdot 10^{k-1} + 1$ distinct k-digit numbers, $0 < r < 9$. Prove that you can choose $r + 1$ numbers out of them, such that any two of the $r + 1$ numbers in any decimal location have distinct digits.

1.5 Mathematical Induction

Principle of Mathematical Induction

If the first person in line is a mathematician, and every mathematician in line is followed by a mathematician, then everyone in line is a mathematician.

More seriously: *given a sequence of statements*

$P_1, P_2, \ldots, P_n, \ldots$

If (1) P_s is true (s is a natural number) and (2) for any natural number $k \geq s, P_k \Rightarrow P_{k+1}$ is true,

then all of the statements in the given sequence beginning with P_s are true.

1.5.1 Prove the following formula for any natural n:

$$1 + 3 + 5 + \cdots + (2n - 1) = n^2$$

Solution

This formula in fact consists of a sequence of statements:

$$\begin{array}{ll} P_1: & 1 = 1^2 \\ P_2: & 1 + 3 = 2^2 \\ P_3: & 1 + 3 + 5 = 3^2 \\ & \vdots \\ P_n: & 1 + 3 + 5 + \ldots + (2n + 1) = n^2 \\ & \vdots \end{array}$$

(1) P_1 is certainly true.
(2) Assume P_k is true, i.e.

$$1 + 3 + \ldots + (2k - 1) = k^2$$

By adding $2k + 1$ to both sides of this equality, we get:

$$1 + 3 + 5 + \cdots + (2k - 1) + (2k + 1) = k^2 + 2k + 1,$$

but

$$k^2 + 2k + 1 = (k + 1)^2,$$

therefore,

$$1 + 3 + 5 + \ldots + (2k + 1) = (k + 1)^2,$$

i.e., P_{k+1} is true.

We proved that the implication $P_k \Rightarrow P_{k+1}$ is true for any natural k.

By Principal of Mathematical Induction, P_n is true for every natural n. The formula is proven.

■

I can't resist to show here one more solution of this problem, a geometric one (please see Figure 1.3).

See:

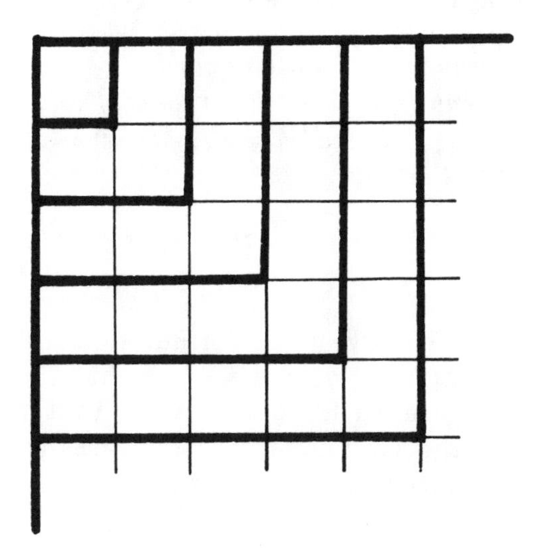

Figure 1.3

I don't present here the solution of the next problem: it is the reader's exercise.

1.5.2 Prove that for any natural n

$$1 + 2 + \ldots + n = \frac{n(n+1)}{2}$$

Please note that condition (2) in the Principle of Mathematical Induction only requires that the implication $P_k \Rightarrow$

P_{k+1} be true. It does not require that P_{k+1} be true. In fact, a sequence $P_1, P_2, \ldots, P_n \ldots$ of all false statements satisfies the condition (2)!

Thus, the condition (1) that requires that P_s be true, is an essential part of the Principle.

Another formulation of the Principle of Mathematical Induction, equivalent to the original one, but yet giving us a somewhat more powerful tool is the following.

Principle of Mathematical Induction II

Given a sequence of statements $P_1, P_2, \ldots, P_n, \ldots$
If (1) P_s is true (s is a natural number) and
(2) $(P_s \wedge P_{s+1} \wedge \ldots \wedge P_k) \Rightarrow P_{k+1}$ is true for any natural $k \geq s$;

then all of the statements in the given sequence beginning with P_s are true.

The difference between this formulation and the original one is in the fact that we can assume that the statements $P_s, P_{s+1}, \ldots, P_k$ are all true, and prove that then P_{k+1} is true as well. And, of course, in some cases this is a more powerful assumption than the assumption that just P_k is true.

1.5.3 (First Annual Southampton Mathematical Olympiad, 1986) Prove that if only two types of coins, a 3-cent coin and a 5-cent coin are minted, then any amount of money greater than 7 cents can be paid in coins.

Solution

(1) $8 = 3+5$, so the required statement is true for 8.

(2) Let $k \geq 8$. Assume that any amount of money greater than 7 and less or equal to k can be paid in coins. We will consider two cases.

(a) If $k-5 \geq 8$, then by assumption $k-5$ can be paid in coins, so can $k+1$:

$$k+1=(k-5)+3+3$$

(b) If $k-5<8$, then $k+1<14$, and we have to check just a few values, namely $k+1=9,10,11,12,13$, but

$$\begin{aligned} 9 &= 3+3+3 \\ 10 &= 5+5 \\ 11 &= 3+3+5 \\ 12 &= 3+3+3+3 \\ 13 &= 3+5+5 \end{aligned}$$

We proved that in either case $k+1$ can be paid in coins.

Thus, any amount of money greater than 7 cents can be paid in coins.

■

The limitation of the method of mathematical induction is in that it can help us verify a hypothesis we have, but offers no help in coming up with a hypothesis. For the latter we have to use our intuition often coupled with experimenting.

1.5.4 We say that several straight lines on the plane are of general position, if no two lines are parallel and no three lines have a point in common. In how many areas do n straight lines of general position partition the plane?

Solution

Let us denote by $S(n)$ the number of areas n straight general lines partition the plane into. Now let us experiment: we draw one line on the plane and get $S(1)=2$; we add one more line to see that $S(2) = 4$; one more line will show that $S(3) = 7$; one more line will demonstrate that $S(4) = 11$ (please see Figures 1.4 and 1.5).

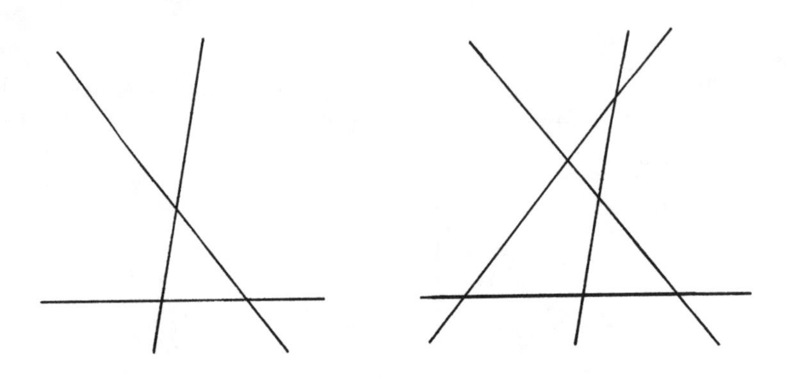

Figure 1.4. $S(3) = 7$ Figure 1.5. $S(4) = 22$

Let us put the data in a table:

Number of Lines n	$S(n)$	Difference $S(n) - S(n-1)$
1	2	
2	4	2
3	7	3
4	11	4

We notice that

$$S(n) = S(n-1) + n$$

This strikingly reminds the relation of the Problem 1.5.2. Remember, $1 + 2 + \cdots + n = \dfrac{n(n+1)}{2}$, i.e. if in that example $S_1(n)$ were to denote $1+2+\cdots+n$, we would get the same relation $S_1(n) = S_1(n-1) + n$.

So let us check the hypothesis $S(n) = \frac{n(n+1)}{2}$

n	$S(n)$	$\frac{n(n+1)}{2}$
1	2	1
2	4	3
3	7	6
4	11	10

Our hypothesis does not work, but we can see from the table above that $S(n)$ and $\frac{n(n+1)}{2}$ differ only by 1!

Thus we conjecture:

$$S(n) = \frac{n(n+1)}{2} + 1$$

As we already know, our conjecture holds for $n = 1$.

Assume that it is true for $n = k$, i.e. k straight lines of general position partition the plane into $S(k) = \frac{k(k+1)}{2} + 1$ areas.

Let $n = k+1$, i.e., we are given $k+1$ straight lines of general position. If we remove one of the lines L, then by inductive assumption the remaining k lines would partition the plane into $S(k) = \frac{k(k+1)}{2} + 1$ areas. Since we have $k+1$ lines of general position, the remaining k lines all intersect the line L; moreover, they intersect L in k different points $a_1, a_2, \ldots, a_k$ (see Figure 1.6).

These k points split the line L into $k+1$ intervals. Each of these intervals splits one area of the partition of the plane by k lines, into two new areas, i.e. instead of $k+1$ old areas we get $2(k+1)$ new areas, i.e.

$$S(k+1) = S(k) + (k+1)$$

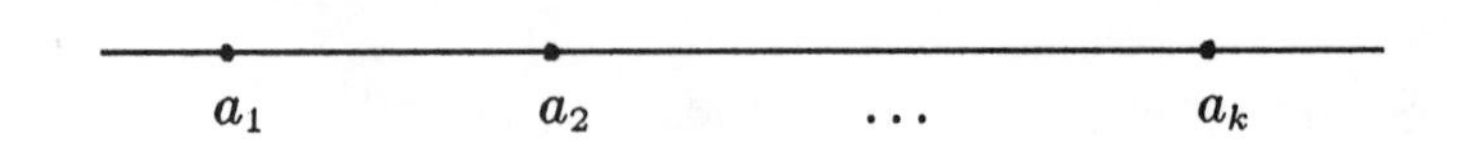

Figure 1.6

therefore,

$$S(k+1) = \frac{k(k+1)}{2} + 1 + (k+1) = \frac{(k+1)(k+2)}{2} + 1$$

In other words, our conjecture holds for $n = k+1$. Thus n straight lines of general position partition the plane into $\frac{n(n+1)}{2} + 1$ areas.

■

Problems

1.5.5 Prove that for any natural n:

$$1^2 + 2^2 + \cdots + n^2 = \frac{n(n+1)(2n+1)}{6}$$

1.5.6 Prove that for any natural n:

$$1^3 + 2^3 + \cdots + n^3 = \frac{n^2(n+1)^2}{4}$$

Note that coupled with Problem 1.5.2, this would give us an interesting corollary:

$$1^3 + 2^3 + \cdots + n^3 = (1 + 2 + \cdots + n)^2$$

1.5.7 Prove that for any natural n:

$$\frac{1}{1 \times 2} + \frac{1}{2 \times 3} + \cdots + \frac{1}{n(n+1)} = \frac{n}{n+1}$$

1.5.8 Prove that for any natural $n, n^3 - n$ is divisible by 6.

1.5.9 Prove that for any natural $n, 11^{n+2} + 12^{2n+1}$ is divisible by 133.

1.5.10 Let n be a natural number. A plane is partitioned into several areas by n straight lines. Prove that the plane can be colored in two colors in such a way that any two areas with a common border (not just one point) are colored in different colors.

Chapter 2
Numbers

2.1 Integers

You remember the definition of a prime number. In Problem 1.3.4 of Section 1.3 we formulated the Fundamental Theorem of Arithmetic. Numerous beautiful results can be presented here, but I will limit myself to problems illustrating some ideas and requiring practically no knowledge of number theory.

2.1.1 Prove that for any integer $n, n^5 - 5n^3 + 4n$ is divisible by 120.

Solution

First of all let us decompose $P(n) = n^5 - 5n^3 + 4n$ as well as 120 into factors:

$$\begin{aligned} P(n) &= n(n^4 - 5n^2 + 4) \\ &= n(n^2 - 1)(n^2 - 4) \\ &= (n-2)(n-1)n(n+1)(n+2) \\ 120 &= 2^3 \cdot 3 \cdot 5 \end{aligned}$$

Since for any integer $n, P(n)$ is a product of five consecutive integers, and one of any five consecutive integers is a multiple of 5, $P(n)$ is divisible by 5.

Similarly, out of any three consecutive integers, one is a multiple of 3, therefore, $P(n)$ is divisible by 3 for any integer n.

Out of any four consecutive integers, one is a multiple of 4, plus one more is even, therefore, $P(n)$ is divisble by $4 \times 2 = 8$ for any integer n.

Due to the Fundamental Theorem of Arithmetic, P(n) is divisible by $2^3 \cdot 3 \cdot 5 = 120$.

■

Given a quadratic equation $ax^2+bx+c=0$. The number $D = b^2 - 4ac$ is called the discriminant of the equation.

2.1.2 Is there an integer x, such that x^2+x+3 is a multiple of 121?

First Solution

Assume that

$$x^2 + x + 3 = 121k,$$

where x and k are integers. We have in fact a quadratic equation in x:

$$x^2 + x + (3 - 121k) = 0$$

In order for a solution to be an integer (remember, the problem asks whether an integer x exists!), the discriminant of the equation has to be a perfect square:

$$4 \times 121k - 11 = n^2,$$

where n is an integer, i.e.,

$$n^2 = 11(4k \times 11 - 1)$$

This means that n^2 is divisible by 11 but is not divisible by 11^2. On the other hand, due to Problem 1.3.4, since n^2 is divisible by 11, n is divisible by 11 as well, which in turn implies that n^2 is divisible by 11^2. Contradiction.

Therefore, there is no integer x such that $x^2 + x + 3$ is a multiple of 121.

■

Second Solution

(x^2+x+3) is divisible by 121 if and only if $4(x^2+x+3)$ is divisible by 121, but

$$4(x^2 + x + 3) = (2x + 1)^2 + 11$$

If for some integers x and k,

$$(2x + 1)^2 + 11 = 11^2 \times k,$$

then

$$(2x + 1)^2 = 11(11k - 1)$$

Just as in the first solution, contradiction is derived from the fact that a square, namely $(2x + 1)^2$, is divisible by 11, and is not divisible by 11^2.

■

2.1.3 Find all integral solutions of the equation

$$x^2 + y^2 + x + y = 3$$

First Solution

Denote $L(x,y) = x^2 + y^2 + x + y = x(x+1) + y(y+1)$. For any integer x, $x(x+1)$ is even as the product of two consecutive integers. Similarly, $y(y+1)$ is even. Thus for any integers x, y, $L(x,y)$ is even and therefore not equal to 3.

The solution set is empty.

■

Second Solution

By multiplying both sides of the given equation by 4, we get:

$$(4x^2 + 4x) + (4y^2 + 4y) = 12,$$

or

$$(2x+1)^2 + (2y+1)^2 = 14$$

On the other hand, a short direct check shows that 14 is not a sum of two squares of integers.

■

Problems

2.1.4 Prove that for any integer $n, n^5 - n$ is divisible by 30.

2.1.5 Prove that for a prime p greater than 3, $p^2 - 1$ is divisible by 24.

2.1.6 Prove that for any prime p and prime q each greater than 3, $p^2 - q^2$ is divisible by 24.

2.1.7 Can $4p+1$ be a prime number, if both p and $2p+1$ are primes, and $p > 3$?

2.1.8 Prove that the remainder upon dividing any prime number by 30 is again a prime.

2.1.9 Solve the following equation in integers:

$$15x^2 - 7y^2 = 9$$

2.1.10 Prove that for any natural $n, 10^n + 18n - 1$ is divisible by 27.

2.2 Rational and Irrational Numbers

We already met rational and irrational numbers in Section 1.3 (Problems 1.3.1, 1.3.3, 1.3.5). As you know, rational numbers are the ones that can be presented in a form m/n, where m and n are integers and $n \neq 0$. But how does one recognize whether a number given as a decimal fraction is rational or irrational? Rational numbers are exactly terminal or infinite repeating decimal fractions.

2.2.1 Prove that the number

$$A = 0.101001000\ldots,$$

where the number of zeros between units increases by one, is irrational.

Solution

Assume that A is a repeating fraction, i.e. after the first k digits, the same sequence of n digits (we'll call it period) is repeating. Since the number of consecutive zeros in the decimal representation of A is increasing, we can find $2n+k$

consecutive zeros, but this implies that all n digits of the period are zeros, therefore in the decimal decomposition of A from some point on we get only zeros.

This contradicts, however, the definition of A, which allows us to find a digit 1 further right than any given digit of the decimal representation of A.

■

2.2.2 Numbers $a, b, \sqrt{a}+\sqrt{b}$ are rational. Prove that numbers $\sqrt{a}$, and $\sqrt{b}$ are rational as well.

Solution

Numbers a and b are rational, therefore $(a+b)$ is rational. Numbers $(a+b)$ and $(\sqrt{a}+\sqrt{b})$ are rational, thus $\sqrt{a}-\sqrt{b} = \dfrac{a+b}{\sqrt{a}+\sqrt{b}}$ is rational. Now,

$$\sqrt{a} = \frac{1}{2}[(\sqrt{a}+\sqrt{b}) + (\sqrt{a}-\sqrt{b})]$$

is rational; so is $\sqrt{b} = (\sqrt{a}+\sqrt{b}) - \sqrt{a}$.

■

2.2.3 Prove that the number $1+\sqrt{5}$ cannot be written as a sum of squares of numbers of the form $a+b\sqrt{5}$ with rational a and b.

Solution

Let us first note that if for integers x_1, x_2, y_1, y_2

$$x_1 + y_1\sqrt{5} = x_2 + y_2\sqrt{5},$$

then $x_1 = x_2$ and $y_1 = y_2$. Indeed, otherwise we would get $x_1 \neq x_2$ *and* $y_1 \neq y_2$, thus $\sqrt{5} = \frac{x_1 - x_2}{y_2 - y_1}$ with an irrational left side and a rational right side of equality.

Now let us assume that

$$1 + \sqrt{5} = (a_1 + b_1\sqrt{5})^2 + (a_2 + b_2\sqrt{5})^2 + \cdots + (a_n + b_n\sqrt{5})^2$$

Due to the uniqueness proven above, we can conclude that then

$$1 - \sqrt{5} = (a_1 - b_1\sqrt{5})^2 + (a_2 - b_2\sqrt{5})^2 + \cdots + (a_n - b_n\sqrt{5})^2$$

But $1 - \sqrt{5} < 0$, and the right side is non-negative as a sum of squares, contradiction.

■

2.2.4 Let p/q, where p and q are integers and their greatest common divisor $\gcd(p, q) = 1$, be a solution of an algebraic equation

$$a_0x^n + a_1x^{n-1} + \cdots + a_n = 0$$

with integral coefficients. Prove that p is a divisor of a_n, and q is a divisor of a_0.

Solution

Essentially we are given the equality

$$a_0\frac{p^n}{q^n} + a_1\frac{p^{n-1}}{q^{n-1}} + \cdots + a_{n-1}\frac{p}{q} + a_n = 0,$$

therefore:

$$a_0p^n = q(-a_1p^{n-1} - \cdots - a_nq^{n-1}),$$

i.e., q is a divisor of $a_0 p^n$. Since $\gcd(p, q) = 1$, it implies that q is a divisor of a_0.

Similarly,

$$a_n q^n = p(-a_{n-1} q^{n-1} - \cdots - a_0 p^{n-1})$$

and thus p is a divisor of a_n.

■

The statement of Problem 2.2.4 has a very important conseqence:

Corollary

Any rational solution of the equation

$$x^n + a_1 x^{n-1} + \cdots + a_n = 0$$

with integral coefficients is an integer (prove it!).

Problems

2.2.5 Number A is given as a decimal fraction:

$$A = 0.10000000001\ldots,$$

where units occupy the first, tenth, hundredth, thousandth, etc., positions after the dot, with zeros everywhere else. Prove that:

(a) A is an irrational number;

(b) A^2 is an irrational number.

2.2.6 Prove that for any integer n,

$$\frac{n}{3} + \frac{n^2}{2} + \frac{n^3}{6}$$

is an integer.

2.2.7 Prove that for any natural number n,

$$\frac{n}{6}+\frac{n^2}{2}+\frac{n^3}{3}$$

is an integer.

2.2.8 Prove that $\sqrt[3]{2}$ cannot be written in the form $p+q\sqrt{r}$ where p, q, r are rational numbers.

2.2.9 Solve Problem 1.5.7 from Section 1.5. without use of mathematical induction, i.e., prove that for any natural n,

$$\frac{1}{1\times 2}+\frac{1}{2\times 3}+\cdots+\frac{1}{n(n+1)}=\frac{n}{n+1}$$

Chapter 3
Algebra

3.1 Proof of Equalities and Inequalities

3.1.1 Prove that for any real numbers a, b, c the sum $a + b + c$ is a divisor of

$$a^3 + b^3 + c^3 - 3abc.$$

Solution

By using twice the equality

$$x^3 + y^3 = (x+y)^3 - 3xy(x+y)$$

we get:

$$\begin{aligned} a^3 + b^3 + c^3 - 3abc &= (a+b)^3 + c^3 - 3ab(a+b) - 3abc \\ &= (a+b+c)^3 - 3(a+b)c(a+b+c)- \\ &\quad 3ab(a+b+c) \\ &= (a+b+c)(a^2+b^2+c^2-ab- \\ &\quad ac-bc) \end{aligned}$$

■

3.1.2 What is wrong with the following "proof" of the inequality $\frac{a+b}{2} \geq \sqrt{ab}$:

$$\begin{array}{rcl} \frac{a+b}{2} & \geq & \sqrt{ab} \\ & \Downarrow & \\ \frac{(a+b)^2}{4} & \geq & ab \\ & \Downarrow & \\ a^2+2ab+b^2 & \geq & 4ab \\ & \Downarrow & \\ (a-b)^2 & \geq & 0 \end{array}$$

The last inequality is true, therefore

$$\frac{a+b}{2} \geq \sqrt{ab}$$

Solution

Something must be wrong because the inequality to be proven is false for, say, negative a and b, and is not defined when one of the numbers a, b is positive and one negative. All of the implications in our chain are true, but the fact that we deduced a true inequality from the one to be proven has nothing to do with proving that inequality. Well, almost nothing. This chain can serve as analysis which can help find a proof, but the proof must be a chain of implications which starts with the inequality(s) known to be true and ending with the required inequality.

■

3.1.3 Prove that for any non-negative a, b,

$$\frac{a+b}{2} \geq \sqrt{ab}$$

First Solution

(a) Analysis:

$$\begin{aligned} \frac{a+b}{2} &\geq \sqrt{ab} \\ &\Downarrow \\ a - 2\sqrt{ab} + b &\geq 0 \\ &\Downarrow \\ (\sqrt{a} - \sqrt{b})^2 &\geq 0 \end{aligned}$$

(b) Proof:

$$\begin{aligned} (\sqrt{a} - \sqrt{b})^2 &\geq 0 \quad \text{for any non-negative } a, b \text{ as a square of a real number} \\ &\Downarrow \\ a - 2\sqrt{ab} + b &\geq 0 \\ &\Downarrow \\ \frac{a+b}{2} &\geq \sqrt{ab} \end{aligned}$$

■

Instead of tracing two chains of implications, we can make sure that every implication in the original chain (analysis) is reversible.

Second Solution

$$\begin{aligned} \frac{a+b}{2} &\geq \sqrt{ab} \\ &\Updownarrow \\ a - 2\sqrt{ab} + b &\geq 0 \\ &\Updownarrow \\ (\sqrt{a} - \sqrt{b})^2 &\geq 0 \end{aligned}$$

which is true for any non-negative a and b.

Please note that the equality is achieved if and only if $a = b$.

■

3.1.4 Prove that for any non-negative x, y, z:

$$x^2 + y^2 + z^2 \geq xy + yz + xz$$

Solution

Using the inequality proven above (Problem 3.1.3) three times, we get:

$$\frac{x^2 + y^2}{2} \geq xy$$

$$\frac{y^2 + z^2}{2} \geq yz$$

$$\frac{x^2 + z^2}{2} \geq xz$$

All there is left to do is to add up these three inequalities.

■

3.1.5 Prove that for any non-negative a, b, c:

$$\frac{a + b + c}{3} \geq \sqrt[3]{abc}$$

Solution

From Problem 3.1.1 we know that:

$$x^3 + y^3 + z^3 - 3xyz = (x + y + z)(x^2 + y^2 + z^2 - xy - yz - xz)$$

Due to the inequality proven above (Problem 3.1.4),

$$x^2 + y^2 + z^2 - xy - yz - xz \geq 0$$

Combining these two facts, we conclude that for any non-negative x, y, z

$$x^3 + y^3 + z^3 - 3xyz \geq 0$$

All there is left to do is to denote $x^3 = a, y^3 = b, y^3 = c$ to get the required

$$\frac{a+b+c}{3} \geq \sqrt[3]{abc}.$$

■

3.1.6 Prove that for any natural n and non-negative $a_1, a_2, \ldots, a_n$:

$$\frac{a_1 + a_2 + \cdots + a_n}{n} \geq \sqrt[n]{a_1 a_2 \ldots a_n}$$

Moreover, the equality is achieved if an only if $a_1 = a_2 = \ldots = a_n$.

Solution

(a) We will first prove by induction on m that this inequality is true for $n = 2^m$. It is true for $m = 1$ (we proved it in Problem 3.1.3).

Assume that the inequality is true for $p = 2^k$, i.e., for any non-negative $a_1, a_2, \ldots, a_p$

$$\frac{a_1 + a_2 + \ldots + a_p}{p} \geq \sqrt[p]{a_1 a_2 \ldots a_p}$$

Now let $a_1, a_2, \ldots, a_{2p}$ be non-negative numbers. By using first the inductive assumption and then the inequality of Problem 3.1.3, we get:

$$\begin{aligned}
&\frac{a_1 + a_2 + \cdots + a_{2p}}{2p} \\
&= \frac{a_1 + a_2}{2} + \frac{a_3 + a_4}{2} + \cdots + \frac{a_{2p-1} + a_{2p}}{2} \\
&\geq \sqrt[p]{\frac{a_1 + a_2}{2} \frac{a_3 + a_4}{2} \cdots \frac{a_{2p-1} + a_{2p}}{2}} \\
&\geq \sqrt[p]{\sqrt{a_1 a_2}\sqrt{a_3 a_4} \cdots \sqrt{a_{2p-1} a_{2p}}} \\
&= \sqrt[2p]{a_1 a_2 \ldots a_{2p}}
\end{aligned}$$

(b) Now we can prove the required inequality for any natural n.

Indeed, if $n = 2^m$ for some m, the inequality is proven in (a); otherwise there exist natural numbers t and m such that $n + t = 2^m$, and for any non-negative $a_1, a_2, \ldots, a_{n+t}$

$$\frac{a_1 + a_2 + \cdots + a_n + a_{n+1} + \cdots + a_{n+t}}{n+t} \geq \sqrt[n+t]{a_1 a_2 \ldots a_n a_{n+1} \ldots a_{n+t}}$$

All there is left to do is to define:

$$a_{n+1} = a_{n+2} = \ldots = a_{n+t} = \frac{a_1 + a_2 + \cdots + a_n}{n}$$

and to plug it in the inequality above:

$$\frac{a_1 + a_2 + \cdots + a_n + t \cdot \dfrac{a_1 + a_2 + \cdots + a_n}{n}}{n+t} \geq \sqrt[n+t]{a_1 a_2 \ldots a_n \left(\frac{a_1 + a_2 + \cdots + a_n}{n}\right)^t}$$

If we denote $\frac{a_1 + a_2 + \cdots + a_n}{n} = A$; $\sqrt[n]{a_1 a_2 \ldots a_n} = B$ the inequality above can be rewritten as follows:

$$\frac{nA + tA}{n+t} \geq \sqrt[n+t]{B^n A^t},$$

i.e.

$$A \geq \sqrt[n+t]{B^n A^t},$$

therefore,

$$A^{n+t} \geq B^n A^t,$$

or

$$A^n \geq B^n,$$

and finally

$$A \geq B,$$

but this is exactly the required inequality!

(c) If $a_1 = a_2 = \ldots = a_n \geq 0$ then we get the equality. Assume now that not all non-negative numbers $a_1, a_2, \ldots, a_n$ are equal. Without loss of generality, we can assume that $a_1 \neq a_2$, then:

$$\frac{a_1 + a_2}{2} > \sqrt{a_1 a_2}, \quad \text{or} \quad \frac{(a_1 + a_2)^2}{2} > a_1 a_2,$$

and we get:

$$\frac{a_1 + a_2 + \cdots + a_n}{n} =$$

$$\frac{\frac{a_1 + a_2}{2} + \frac{a_1 + a_2}{2} + a_3 + \cdots + a_n}{n} \geq$$

$$\sqrt[n]{\left(\frac{a_1 + a_2}{2}\right)^2 a_3 \ldots a_n} >$$

$$\sqrt[n]{a_1 a_2 a_3 \cdots a_n},$$

thus the equality takes place if and only if $a_1 = a_2 = \ldots = a_n$.

■

3.1.7 Prove that for any natural $n \geq 2$

$$\frac{1}{2^2} + \frac{1}{3^2} + \cdots + \frac{1}{n^2} < 1$$

Solution

Clearly,

$$\frac{1}{2^2} + \frac{1}{3^2} + \cdots + \frac{1}{n^2} < \frac{1}{1 \cdot 2} + \frac{1}{2 \cdot 3} + \cdots + \frac{1}{(n-1)n}$$

Now, if you solved Problem 1.5.7 or 2.2.9, you know that:

$$\frac{1}{1 \cdot 2} + \frac{1}{2 \cdot 3} + \cdots + \frac{1}{(n-1)n} < \frac{n-1}{n}$$

therefore,

$$\frac{1}{2^2} + \frac{1}{3^2} + \cdots + \frac{1}{n^2} < \frac{n-1}{n} < 1$$

But what if you could not solve Problem 2.2.9? Here is a solution for you:

$$\frac{1}{1 \cdot 2} + \frac{1}{2 \cdot 3} + \cdots + \frac{1}{(n-1)n} =$$

$$\left(\frac{1}{1} - \frac{1}{2}\right) + \left(\frac{1}{2} - \frac{1}{3}\right) + \cdots + \left(\frac{1}{n-1} - \frac{1}{n}\right)$$

As you can see, all the terms, except for the first and the last ones, cancel out, and we get the required:

$$1 - \frac{1}{n} = \frac{n-1}{n}$$

■

Problems

3.1.8 Prove that for any positive a and b

$$\frac{2}{\frac{1}{a}+\frac{1}{b}} \leq \sqrt{ab} \leq \frac{a+b}{2} \leq \frac{\sqrt{a^2+b^2}}{2}$$

Moreover, equalities take place if and only if $a = b$.

3.1.9 Prove that for any non-negative a and b

$$\frac{a+b}{2} \leq \sqrt[3]{\frac{a^3+b^3}{2}}$$

3.1.10 Prove that for any non-negative x, y, z

$$(x+y)(y+z)(x+z) \geq 8xyz$$

3.1.11 Prove that for any $a_1, a_2, \ldots, a_n$

$$\frac{a_1+a_2+\cdots+a_n}{n} \leq \sqrt{\frac{{a_1}^2+{a_2}^2+\cdots+{a_n}^2}{n}}$$

3.1.12 Prove that for any natural n,

$$\frac{1}{3^2}+\frac{1}{5^2}+\cdots+\frac{1}{(2n+1)^2} < \frac{1}{4}$$

3.1.13 Prove a better estimate than the one we proved in Problem 3.1.7: for any natural $n \geq 2$,

$$\frac{1}{2^2}+\frac{1}{3^2}+\cdots+\frac{1}{n^2} < \frac{2}{3}$$

3.2 Equations, Inequalities, Their Systems, and How to Solve Them

The wealth of material available for this section is so great, that it alone can fill up a book, or even books. Without any claim on completeness, I just want to present here a few ideas. Unless otherwise stated, we solve equations and inequalities in the set of real numbers.

3.2.1 Solve the following equation:

$$(x^2 - x + 1)(x^2 - x + 2) = 12$$

Solution

Denote $x^2 - x + 1 = y$ and express the given equation in terms of y:

$$y(y+1) = 12,$$

i.e.

$$y^2 + y - 12 = 0$$

Now we can solve this quadratic equation:

$$y_1 = -4 \quad \Big| \quad y_2 = 3$$

Rewriting it back in terms of x, we get:

	$x^2 - x + 1 = -4$	$x^2 - x + 1 = 3$
or	$x^2 - x + 5 = 0$	$x^2 - x - 2 = 0$
	no solutions in real numbers	$x_1 = -1$ or $x_2 = 2$

Thus the solution set is $\{-1, 2\}$.

■

The reader is probably familiar with Viète Theorem: *The solution set $\{x_1, y_1\}$ of the quadratic equation*

$$z^2 + pz + q = 0$$

satisfies the following system of equalities:

$$\begin{cases} x_1 + y_1 = -p \\ x_1 y_1 = q \end{cases}$$

It is very helpful sometimes to know (prove it!) that the converse of Viète Theorem is true too:

If the ordered pair (x_1, y_1) is a solution of the system

$$\begin{cases} x + y = -p \\ xy = q \end{cases}$$

then the set $\{x_1, y_1\}$ is the solution set of the equation

$$z^2 + pz + q = 0$$

3.2.2 Solve the following system of equations:

$$\begin{cases} x^3 + y^3 = -7 \\ xy = -2 \end{cases}$$

Solution

Denote

$$\begin{cases} x + y = -p \\ xy = q \end{cases}$$

If we can find p and q, then by the converse of Viète Theorem, $\{x, y\}$ will be the solution set of the equation

$$z^2 + pz + q = 0$$

Let us therefore rewrite the given system in terms of p and q. Due to the fact that $x^3+y^3=(x+y)^3-3xy(x+y)$, it is easy to do:

$$\begin{cases}(-p)^3-3q(-p)=-7\\ q=-2\end{cases}$$

i.e.,

$$\begin{cases}p^3-3pq=7\\ q=-2\end{cases}$$

By substituting -2 for q in the first equation, we get

$$p^3+6p-7=0$$

i.e.

$$(p-1)(p^2+p+7)=0$$

$$\begin{array}{l|l} p-1=0 & p^2+p+7=0\\ p=1 & \text{no real solutions}\end{array}$$

We have $\begin{cases}p=1\\ q=-2\end{cases}$, therefore $\{x,y\}$ is the solution set of the equation $z^2+z-2=0$. But $z_1=1; z_2=-2$, therefore the original system has two solutions:

$$\begin{cases}x=1\\ y=-2\end{cases};\qquad \begin{cases}x=-2\\ y=1\end{cases}$$

■

Quite often in order to solve a given system of equations, we replace it by one equation. It is much more surprising that *sometimes replacing the given equation by a system of equations proves to be very productive too.*

3.2.3 Solve the following equation:

$$x^2+\left(\frac{x}{x+1}\right)^2=1$$

Solution

Straight forward simplification will bring you (check!) to a pretty hopeless equation:

$$x^4 - 2x^3 - x - 1 = 0$$

(Of course, there is a formula for solving the fourth degree equations, but I have yet to meet a person who remembers it!)

Denote $\frac{x}{x+1} = y$. Our equation is equivalent to the following system of equations:

$$\begin{cases} x^2 + y^2 = 1 \\ \frac{x}{x+1} = y \end{cases}$$

or

$$\begin{cases} x^2 + y^2 = 1 \\ x - y = xy \end{cases}$$

Denote

$$\begin{cases} x + (-y) = -p \\ x(-y) = q \end{cases}$$

Since $x^2 + y^2 = [x + (-y)]^2 - 2x(-y)$, this system can be rewritten in terms of p and q as follows:

$$\begin{cases} (-p)^2 - 2q = 1 \\ -p = -q \end{cases}$$

Our goal now is to find p and q, and then to use the converse of Viète Theorem to find x and $-y$.

We have:

$$\begin{cases} p^2 - 2q = 1 \\ p = q \end{cases};$$

therefore,

$$p^2 - 2p - 1 = 0;$$

thus,

$$\begin{cases} p_1 = 1+\sqrt{2} \\ q_1 = 1+\sqrt{2} \end{cases}; \qquad \begin{cases} p_2 = 1-\sqrt{2} \\ q_2 = 1-\sqrt{2} \end{cases}$$

$\{x, -y\}$ is the solution set of the equation

$$z^2 + pz + q = 0$$

Case 1: $p = q = 1 + \sqrt{2}$

$$z^2 + (1+\sqrt{2})z + (1+\sqrt{2}) = 0$$

Since the discriminant

$$D = (1+\sqrt{2})^2 - 4(1+\sqrt{2}) < 0,$$

there is no real solutions.
Case 2: $p = q = 1 - \sqrt{2}$

$$z^2 + (1-\sqrt{2})z + (1-\sqrt{2}) = 0$$

$$z_{1.2} = \frac{\sqrt{2} - 1 \pm \sqrt{2\sqrt{2}-1}}{2}$$

Therefore, we get two solutions:

$$\begin{cases} x = z_1 \\ -y = z_2 \end{cases}; \qquad \begin{cases} x = z_2 \\ -y = z_1 \end{cases}$$

Since we are not really interested in solutions for y (after all, y was not a part of the original problem, it was our creation!), we finally get:

$$x_1 = \tfrac{1}{2}(\sqrt{2} - 1 + \sqrt{2\sqrt{2}-1});$$
$$x_2 = \tfrac{1}{2}(\sqrt{2} - 1 - \sqrt{2\sqrt{2}-1})$$

■

3.2.4 Solve the following equation:

$$x^{x^2-5x+6} = 1$$

Solution

By taking the natural logarithm of both sides of the equation, we get:

$$(x^2 - 5x + 6)\ln|x| = 0$$

(Please note that in order not to lose any solutions, we must use the absolute value sign. As a result, we might gain some extraneous solutions, therefore all solutions x such that $x < 0$, will have to be checked.)

$x^2 - 5x + 6 = 0$	$\ln|x| = 0$
$x_1 = 2;$ $x_2 = 3;$	$x_3 = 1;$ $x_4 = -1$

The check shows that $x_4 = -1$ is a solution of this equation, as well as $x = 1, 2, 3$.

■

3.2.5 Solve the following equation:

$$\sqrt{x+1} - \sqrt{x-1} = 1$$

Solution

By multiplying both sides of the given equation by $\sqrt{x+1} + \sqrt{x-1}$, we get:

$$\sqrt{x+1} + \sqrt{x-1} = 2$$

Now we add this equation with the original one:

$$2\sqrt{x+1} = 3,$$

therefore,

$$x + 1 = \frac{9}{4},$$

or

$$x = \frac{5}{4}.$$

■

3.2.6 Solve the following equation:

$$\sqrt[3]{1+\sqrt{x}} + \sqrt[3]{1-\sqrt{x}} = \sqrt[3]{5}$$

Solution

Keeping in mind that

$$(a+b)^3 = a^3 + b^3 + 3ab(a+b),$$

we compute the cubes of both sides of the given equation:

$$2 + 3\sqrt[3]{1-x}\left(\sqrt[3]{1+\sqrt{x}} + \sqrt[3]{1-\sqrt{x}}\right) = 5$$

But $\sqrt[3]{1+\sqrt{x}} + \sqrt[3]{1-\sqrt{x}} = \sqrt[3]{5}$; therefore, we get:

$$2 + 3\sqrt[3]{1-x} \cdot \sqrt[3]{5} = 5,$$

i.e.

$$\sqrt[3]{1-x} = \frac{1}{\sqrt[3]{5}},$$

$$1 - x = \frac{1}{5},$$

$$x = \frac{4}{5}$$

■

3.2.7 Solve the following inequality:

$$\sqrt{\frac{x^3+8}{x}} > x - 2$$

Solution

First we determine the domain of the given inequality:

$$\begin{cases} x^3 + 8 \geq 0 \\ x > 0 \end{cases} \Rightarrow \begin{cases} x \geq -2 \\ x > 0 \end{cases} \Rightarrow x > 0$$

or

$$\begin{cases} x^3 + 8 \leq 0 \\ x < 0 \end{cases} \Rightarrow \begin{cases} x \leq -2 \\ x < 0 \end{cases} \Rightarrow x \leq -2$$

Thus the domain is $(-\infty, -2] \cup (0, +\infty)$. We will consider two cases:

(a) $x - 2 < 0$, or $x < 2$. In this case any x from the domain will be a solution of the inequality because its left side is non-negative whereas its right side is negative.

(b) $x - 2 \geq 0$, or $x \geq 2$. In this case our inequality is equivalent to the following:

$$\frac{x^3 - 8}{x} > (x - 2)^2$$

Since in this case $x > 0$, we get

$$x^3 + 8 > x(x - 2)^2,$$

i.e.

$$x^3 + 8 > x^3 - 4x^2 + 4x,$$
$$4x^2 - 4x + 8 > 0,$$
$$(2x - 1)^2 + 7 > 0,$$

which is true for any x from the domain.

Thus we finally get the union of two intervals: $(-\infty, -2] \cup (0, +\infty)$

■

3.2.8 Solve the following inequality:

$$x^5 - 5x^3 + 4x > 0$$

Solution

As we discovered in Problem 2.1.1, $P(x) = x^5 - 5x^3 + 4x = (x-2)(x-1)x(x+1)(x+2)$.

Let us now draw the number line and plot on it the roots of $P(x)$:

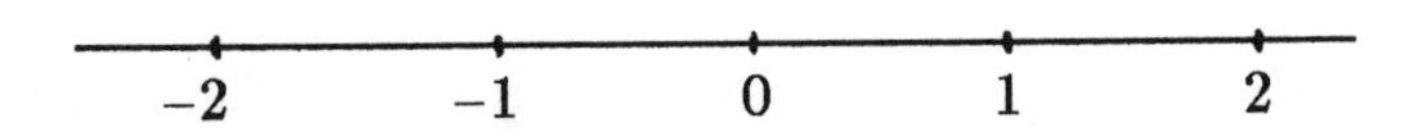

$P(x)$ is the product of five factors, all of which are positive when $x > 2$. Then as we move left along the number line, exactly one factor changes the sign when we pass every point plotted on the number line above.

This observation gives away the solution of the inequality: all we have to do is to draw a sine-type curve line through all of the plotted points, which begins above the number line where $x > 2$:

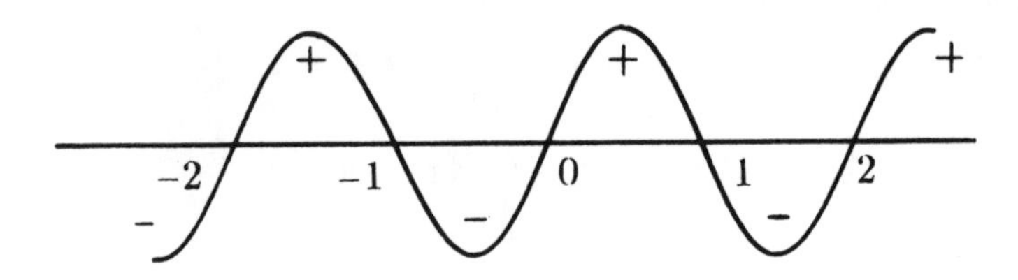

Since we are interested for which $x, P(x) > 0$, the answer is

$$(-2,-1) \cup (0,1) \cup (2,\infty)$$

■

3.2.9 Prove that if (x, y, z) is a solution of the system of equations

$$\begin{cases} x+y+z=a \\ \dfrac{1}{x}+\dfrac{1}{y}+\dfrac{1}{z}=\dfrac{1}{a} \end{cases}$$

then at least one of the numbers x, y, z is equal to a.

Solution

Let (x, y, z) be a solution of the given system, then from the second equation

$$\frac{1}{x}+\frac{1}{y}=\frac{1}{a}-\frac{1}{z}$$

i.e.

$$\frac{x+y}{xy}=-\frac{a-z}{az}$$

But from the first equation $x+y=a-z$, therefore

$$xy(a-z)=-az(a-z)$$

Assume that $z \neq a$, then by dividing by $(a-z)$, we get $xy = -az$. So we obtained above the following two equalities:

$$\begin{cases} x+y=a-z \\ xy=-az \end{cases}$$

By the converse of Viète Theorem, $\{x, y\}$ is the solution set of the following quadratic equation:

$$v^2-(a-z)v-az=0$$

i.e.

$$(v-a)(v+z)=0$$

Since one of the solutions of this equation is $v_1=a, x$ or y must be equal to v_1, i.e. to a.

■

3.2.10 Solve the following equation:

$$x^2 + \frac{1}{x^2} = 2^{1-y^2}$$

Solution

As we know from Section 3.1, for any non-negative a, b,

$$a + b \geq 2\sqrt{ab}$$

therefore,

$$x^2 + \frac{1}{x^2} \geq 2\sqrt{x^2 \frac{1}{x^2}} = 2$$

So the minimum of the left side of the equation is 2. On the other hand, the maximum of the right side is 2, therefore the given equation is equivalent to the system:

$$\begin{cases} x^2 + \frac{1}{x^2} = 2 \\ 2^{1-y^2} = 2 \end{cases}$$

By solving the equations of the system separately, we get $x = \pm 1$; $y = \pm 1$; so we have four solutions:

$$\begin{cases} x_1 = 1 \\ y_1 = 1 \end{cases}; \quad \begin{cases} x_2 = 1 \\ y_2 = -1 \end{cases}; \quad \begin{cases} x_3 = -1 \\ y_3 = 1 \end{cases}; \quad \begin{cases} x_4 = -1 \\ y_4 = -1 \end{cases}$$

■

Problems

3.2.11 Solve the following equation:

$$\frac{x^2}{3} + \frac{48}{x^2} = 10\left(\frac{x}{3} - \frac{4}{x}\right)$$

3.2.12 Solve the following system of equations:

$$\begin{cases} 4x^2 + 9y^2 = 10xy + 12 \\ 2x + 3y = 2xy \end{cases}$$

3.2.13 Solve the following equation:

$$x^x = x$$

3.2.14 Solve the following equation:

$$\sqrt{x^2 - 4x + 4} = -(x - 2)$$

3.2.15 Solve the following equation:

$$9 - x^2 = 2x(\sqrt{10 - x^2} - 1)$$

3.2.16 Solve the following equation:

$$\sqrt{x + 3 - 4\sqrt{x - 1}} + \sqrt{x + 8 - 6\sqrt{x - 1}} = 1$$

3.2.17 Solve the following inequality:

$$\frac{x^2 - 7x + 12}{x^2 + x - 20} \geq 0$$

3.2.18 Solve the following inequality:

$$\frac{(x + 1)(x + 3)^2(x + 5)^3}{(x + 2)^3(x + 4)^2(x + 6)} < 0$$

3.2.19 (Moscow State University, Entrance Examination, 1966) Find all the values of a, such that the system of equations in x and y

$$\begin{cases} 2^{|x|} + |x| = y + x^2 + a \\ x^2 + y^2 = 1 \end{cases}$$

has exactly one solution.

Chapter 4
Geometry

In this chapter we will use the following common notations: $|AB|$ for the length of a segment $\overline{AB}$; $m(\alpha)$ for the the measure of an angle α; $S(ABC)$ for the area of a triangle ABC.

4.1 Loci

We start with the four familiar loci on the plane.

4.1.1 Given a point O and a positive real number r. The locus of all points P on the distance r from O is the circle of radius r with the center in O (prove it!).

4.1.2 The locus of all points equidistant from two distinct points A and B is the perpendicular bisector of the segment $\overline{AB}$. (See Problem 1.3.6).

4.1.3 The locus of all points equidistant from two given intersecting straight lines, is a pair of perpendicular (to each other) straight lines which bisect all four angles between the given lines (prove it!).

Let F be a geometric figure and P a point. We say that F is seen from P at the angle α, if α is the angle between two tangent rays drawn from P to F.

4.1.4 Given a segment $\overline{AB}$ and an angle α, $0 < \alpha < \pi$. The locus of all points P from which the segment $\overline{AB}$ is seen at the angle α (i.e., $APB = \alpha$) is a pair of arcs symmetric with respect to the straight line AB, excluding the arcs' endpoints A and B.

Solution

To begin with let us consider only half a plane above the straight line AB. We can draw an isosceles triangle AOB ($|AO| = |OB|$) with $\angle AOB = 2\alpha$, (Figure 4.1), and the circle of radius $|OA|$ with the center in O.

For any point P on the arc AxB, the angle APB is equal to α, because it is measured by one half of the arc AyB (check your geometry textbook for this theorem, but don't close your textbook yet!).

If we take any point P outside of the circle (Figure 4.2.), then $\angle APB$ is less than α, because it is measured by one half the difference of arcs AyB and CzD (testbook!).

If we take any point P inside the circle (Figure 4.3), then $\angle APB$ is greater than α, because it is measured by one half the sum of arcs AyB and CxD (now you can close your textbook!).

Finally, "the story" in half a plane below the line AB is symmetric to the one above AB, therefore we get two symmetric arcs AxB and AvB, excluding the points A and B, of course (Figure 4.4).

■

4.1.5 Given a point O and a straight line L. Find the locus S of all second endpoints P of segments for which one endpoint lies on L, and the midpoint is at the point O.

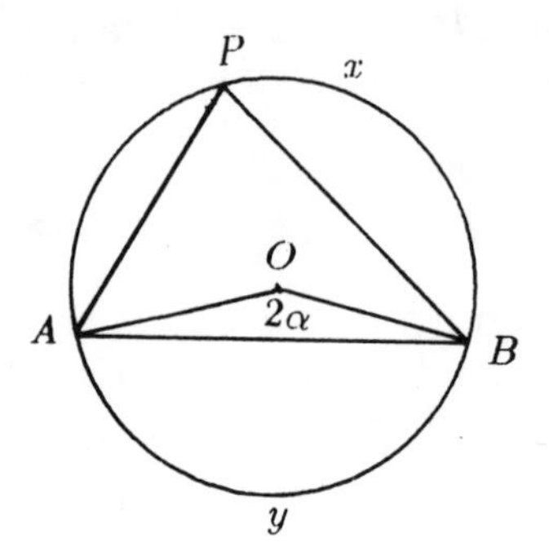

Figure 4.1

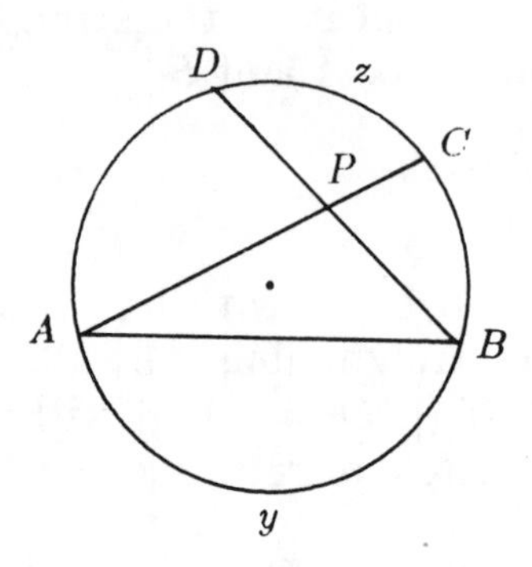

Figure 4.3

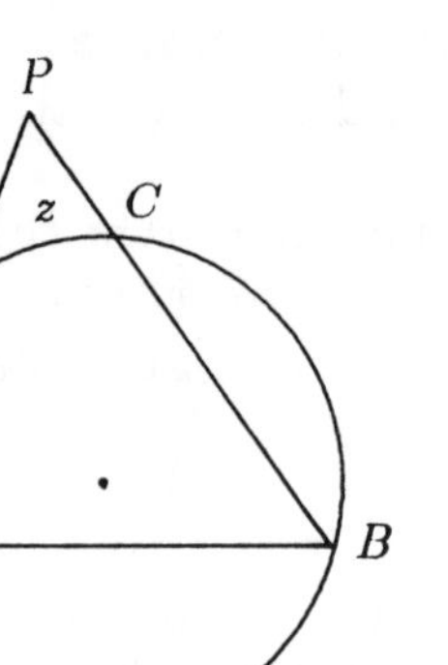

Figure 4.2

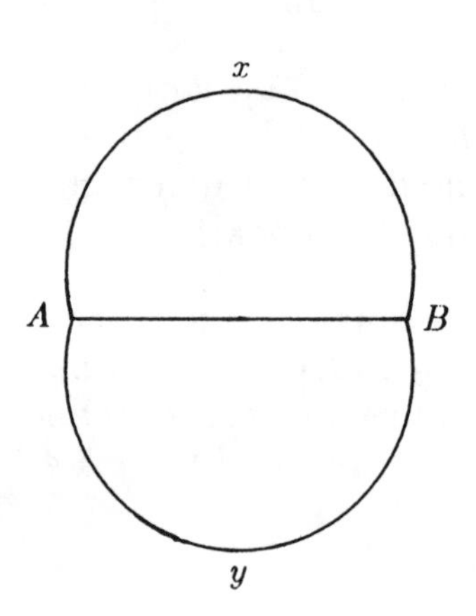

Figure 4.4

Solution

If a point P' lies on L, and O is the midpoint of $\overline{P'P}$, then the point P is the image of the point P' under rotation by π about the point O. Therefore, the set S is the result of rotation of L by π about O.

Of course, S is the straight line parallel to L on the same distance from O as L (see Figure 4.5).

■

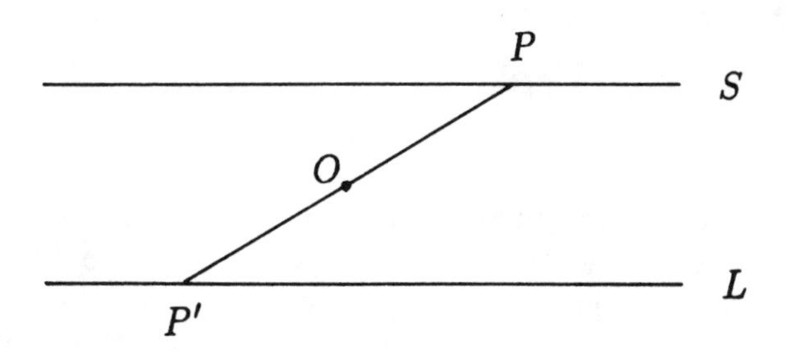

Figure 4.5

Problems

4.1.6 Find the locus of all points equidistant from two given parallel straight lines.

4.1.7 Find the locus of all points P, which are on the given distance r ($r > 0$) from the given straight line L.

4.1.8 Given a segment $\overline{AB}$, a circle X, and angles α and β. Find the locus S of all points P from which the segment $\overline{AB}$ is seen at the angle α and the circle X is seen at the angle β. How many elements can be in the set S?

4.1.9 Given a square X and an angle $0 < \alpha < \frac{\pi}{2}$, find the locus S of all points P from which the square X is seen at the angle α. Solve the same problem, if $\frac{\pi}{2} \leq \alpha < \pi$.

4.1.10 Given the segments $\overline{AB}$ and $\overline{XY}$ of lengths $|AB| = a$ and $|XY| = b$, $a \leq b$. Find the locus of all points P, such that $|PX|^2 - |PY|^2 = a^2$.

4.1.11 Prove that three perpendicular bisectors of the sides of an arbitrary triangle intersect in one point.

4.2 Symmetry and Other Transformations

4.2.1 People living in the neighborhood A and working at the company B (see the map in Figure 4.6) are to drive their children to school on their way to work. Where on highway L should they build the school S to minimize their driving? (When the site S for the school is chosen, the roads $\overline{AS}$ and $\overline{SB}$ are going to be built.)

Analysis

Let A' be the symmetric image of A with respect to L, then $|AS'| + |S'B| = |A'S'| + |S'B|$ (see Figure 4.7). But out of all the broken lines $A'S'B$, the shortest is the segment of the straight line $A'B$.

Solution

Draw the symmetric image A' of A with respect to L, and the straight line $A'B$. The intersection S of $A'B$ and L is the required site for the school.

■

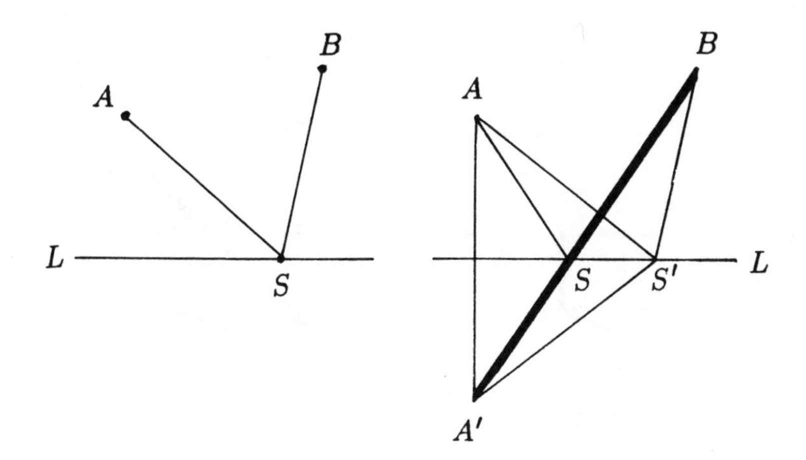

Figure 4.6 The Map Figure 4.7

4.2.2 The river has straight parallel sides, the cities A and B lie on opposite sides of the river (see the map in Figure 4.8). Where shall we build a bridge in order to minimize the travelling distance between A and B (a bridge, of course, must be perpendicular to the sides of the river)?

Analysis

Let B be the result of the translation of B by the width W of the river in the direction perpendicular to the river and toward it (see Figure 4.9). Then

$$|AT'| + |T'S'| + |S'B| = (|AT'| + |T'B'|) + W,$$

but among all the broken lines $AT'B'$ the shortest is the segment of the straight line AB'.

Solution

Draw the image B' of B under the translation by W in the direction perpendicular to the river and toward it, and connect A and B' by the straight line.

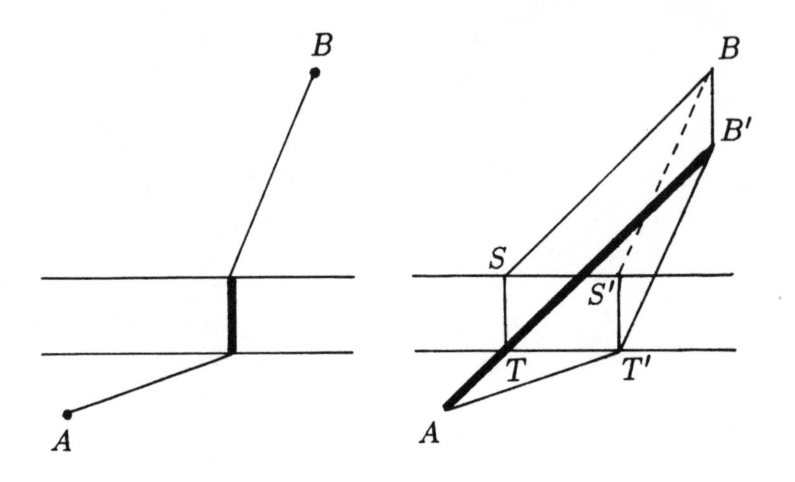

Figure 4.8. The Map Figure 4.9.

The intersection T of AB' with the lower side of the river is the place to build the bridge.

■

4.2.3 Two circles have exactly two points A and B in common. Find a straight line L through A, such that the circles cut out of L chords of equal length. How many solutions can the problem have?

Analysis

At least one solution: AB.

Assume that the line L is not through B, and $|AC| = |AD|$ (see Figure 4.10).

Since the points D and C are symmetric with respect to A, the symmetric image R_1' of the given circle R_1 through D must go through C.

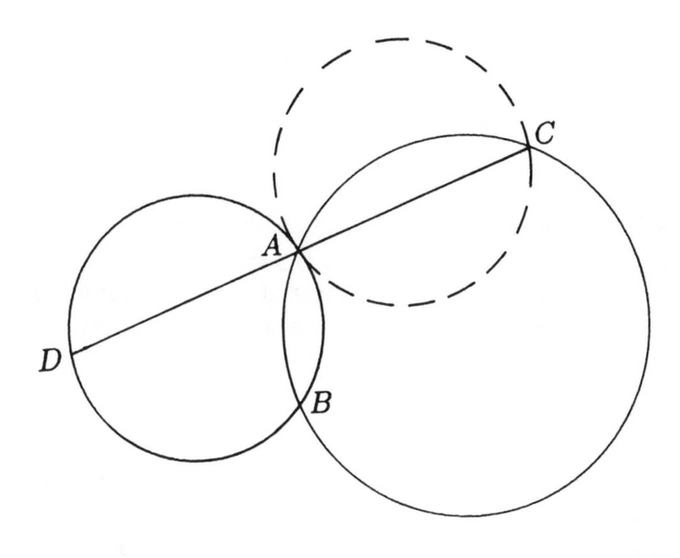

Figure 4.10

Solution

Draw the symmetric image R_1', of the given circle R_1. Every intersection C of R_1' with the other given circle R_2 determines the required straight line CA. The proof is obvious and essentially repeats the analysis.

Research

Since the point A is on R_2, and the circle R_1 has a point X inside R_2 (remember, R_1 and R_2 have exactly two distinct points in common!), the circle R_1' will have a point X' outside of R_2 (namely, the symmetric image X' of X with respect to A). Therefore, R_1' and R_2 have exactly one more point of intersection, in addition to A, and thus the problem always has exactly two solutions (remember, the first solution was AB).

■

We have no doubts that our readers have a good knowledge of some frequently used types of transformations: translation, rotation, central symmetry, line symmetry. We would like to say a few words here about another very important type of transformation: homothety.

Homothety

Given a point O and a non-zero number k. Homothetic transformation, or homothety H with the center of homothety O and coefficient of homothety k is the transformation which maps every point P into the point $P_1 = H(P)$ on the straight line OP, such that

$$\frac{|\overrightarrow{OP_1}|}{|\overrightarrow{OP}|} = k$$

Please note that unlike the case with the length of a segment, the direction of vectors $\overrightarrow{OP}_1$ and $\overrightarrow{OP}$ is taken into account: the measure $|\overrightarrow{a}|$ of a vector $\overrightarrow{a}$ is equal to the length $|a|$ of the corresponding segment $\overline{a}$ if the direction of $\overrightarrow{a}$ is the same as of a unit vector, and is equal to $-|a|$ otherwise. A negative coefficient of homothety k would imply that the points P_1 and P are on opposite sides of the center of homothety O:

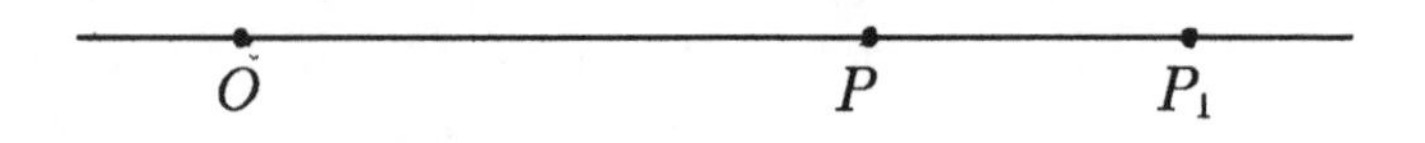

Figure 4.11. Homothety with $k = \frac{3}{2}$

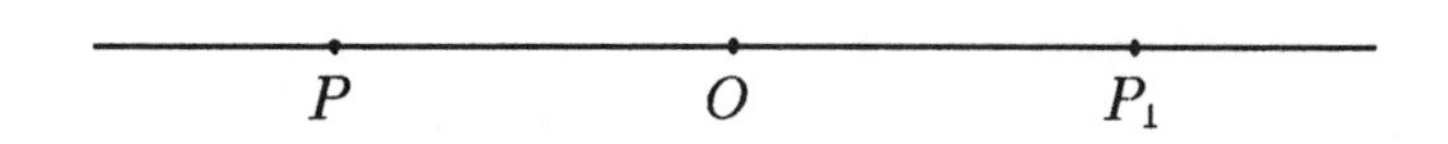

Figure 4.12 Homothety with $k = -1$

The homothetic image $F_1 = H(F)$ of a geometric figure F under homothety H is the geometric figure consisting of homothetic images $P_1 = H(P)$ of all points P which make up the given figure F.

Please note (and prove!) that:

(a) homothetic image $I_1 = H(I)$ of a segment I is the segment I_1 parallel to I such that $\frac{|I_1|}{|I|} = k$, where k is the coefficient of homothety;

(b) two homothetic polygons $H(P)$ and P are similar;

(c) any two circles K_1, K_2 are homothetic (i.e. there exists a homothety H, such that $K_2 = H(K_1)$);

(d) symmetry with respect to a point is a particular case of homothety with the coefficient of homothety $k = -1$.

4.2.4 Inscribe a square in a given acute triangle.

Analysis

Let the square $MNPQ$ be inscribed in the triangle ABC. Two points of the square, say P, Q, must lie on the same side of the triangle, say $\overline{AC}$, with two others one per side of the triangle (see Figure 4.13).

If we waive the requirement, that the vertex N of the square be on $\overline{BC}$, we will get many (in fact infinitely many) squares satisfying all other conditions (i.e., $\overline{PQ}$ to lie on $\overline{AC}$, and M on $\overline{AB}$). Let us take one of them, $M_1N_1P_1Q_1$, then the squares $M_1N_1P_1Q_1$ and MNPQ are homothetic!

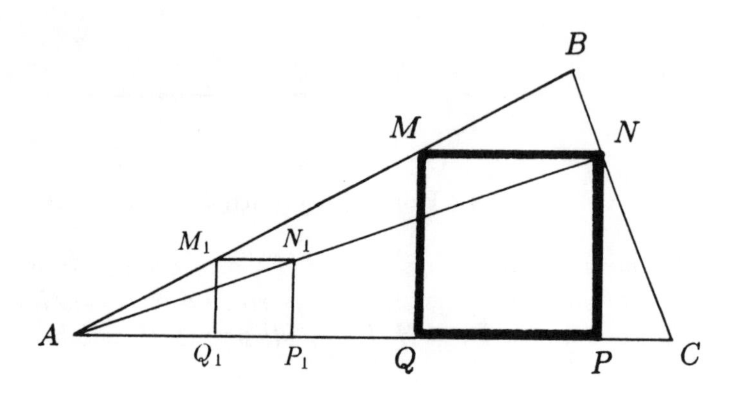

Figure 4.13

Construction

(1) Construct a square $M_1N_1P_1Q_1$ with $\overline{P_1Q_1}$ on $\overline{AC}$ and M_1 on $\overline{AB}$ (pick a point M_1 on $\overline{AB}$, draw $\overline{M_1Q_1} \perp AC$, mark P_1 such that $|Q_1P_1| = |M_1Q_1|$, draw the perpendicular to $\overline{AC}$ at P_1, and the perpendicular to $\overline{M_1Q_1}$ at M_1 with their intersection determining N_1).

(2) Draw the straight line through A and N_1. Denote by N the intersection of AN_1 and BC.

(3) Homothety H with the center in A and the coefficient $k = \dfrac{|AN|}{|AN_1|}$ concludes the construction:

$$H(M_1N_1P_1Q_1) = MNPQ$$

Proof

The homothetic image of a square is a square. $N = H(N_1)$ was chosen to lie on $\overline{BC}$. $M = H(M_1)$ lies on the straight line AM_1, similarly $P = H(P_1)$ and $Q = H(Q_1)$ lie on AC. Finally, since $NM \parallel AC$ and N is an inside point of $\overline{BC}$, M is an inside point of the segment $\overline{AB}$. Similarly you can show that P and Q are inside points of the segment $\overline{AC}$.

Research

Uniqueness and existence of the point N in our construction gives us exactly one solution under the assumption that two vertices of the square lie on $\overline{AC}$. We can construct another solution with two vertices of the square on $\overline{AB}$, and the third one with two vertices of the square on $\overline{BC}$.

Thus the problem always has exactly three solutions.

■

Problems

4.2.5 Where would you build two bridges over the two sleeves of a river with parallel straight sides (see the map in Figure 4.14) to minimize the length of the path between the cities A and B? (Bridges have to be perpendicular to the sides of the sleeves of the river.)

4.2.6 Given three parallel straight lines L_1, L_2, L_3. Find an equilateral triangle ABC with A on L_1, B on L_2, and C on L_3.

4.2.7 Given three concentric circles K_1, K_2, K_3. Find an equilateral triangle ABC with A on K_1, B on K_2, and C on K_3.

4.2.8 Given a point A and a circle K with the center O. We connect A with an arbitrary point B of the circle K. Find the locus of a point P of the intersection of AB with the bisector of the angle AOB.

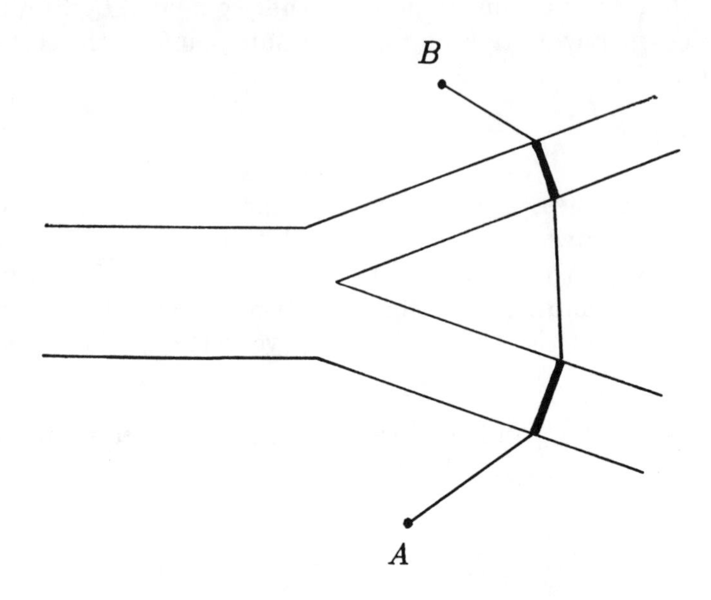

Figure 4.14 The Map

4.3 Proofs in Geometry

4.3.1 Prove that if a and b are two sides of a triangle and m_c is the median drawn to the third side, then

$$|m_c| < \frac{|a| + |b|}{2}$$

Solution

Let $|BO| = |OA|$ (Figure 4.15). We extend CO and mark D, such that $|CO| = |OD|$.

Then the quadrilateral $ACBD$ is a parallelogram, therefore $|BD| = |CA| = |b|$. Finally in the triangle CBD

$$|CD| < |CB| + |BD|,$$

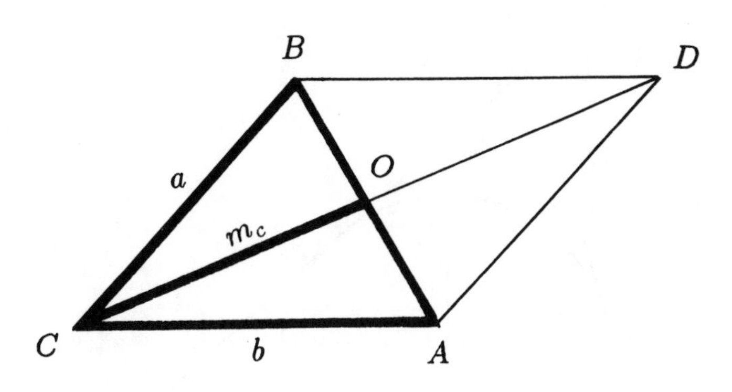

Figure 4.15

i.e.,

$$2|m_c| < |a| + |b|,$$

therefore

$$m_c < \frac{|a| + |b|}{2}$$

■

4.3.2 A circle is inscribed in a triangle ABC. $\overline{MN}$ is the diameter perpendicular to the base $\overline{AC}$. Let L be the intersection of BM with AC (see Figure 4.16). Prove that $|AN| = |LC|$.

We will show here two solutions: a geometric solution and an algebraic one.

First Solution

Draw the tangent line EF to the circle at the point M. $EF \parallel AC$. Consider the homothetic transformation T with

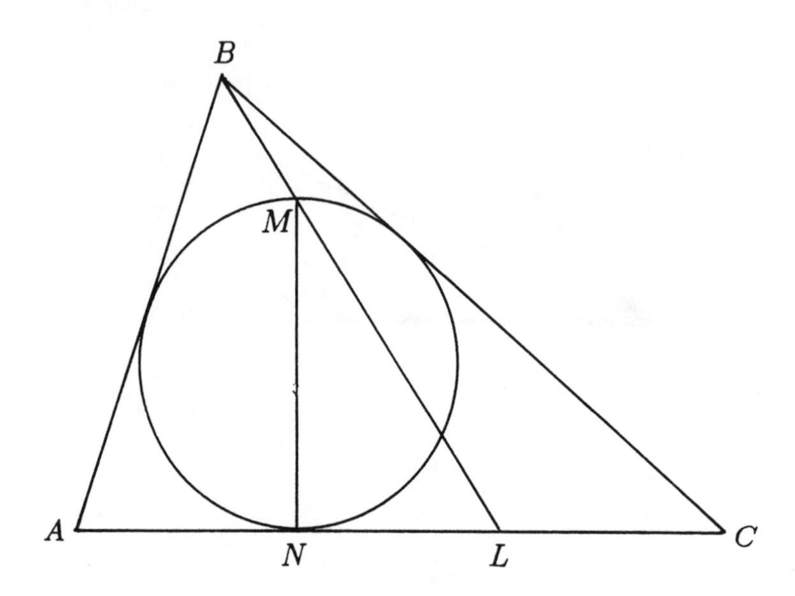

Figure 4.16

the center in B mapping M into L. T will map $\overline{EF}$ onto $\overline{AC}$, and the given circle into the circle tangent to BA, BC, and AC outside of the triangle ABC (see Figure 4.17).

Denote the tangent points of the circles and BA, BC by X, Y, Z, and W respectively.

Since the lengths of the segments of two tangent lines drawn from a point to a circle are equal, we get:

$$\begin{array}{r}|BY| = |BW| \\ -|BX| = |BZ| \\ \hline |XY| = |ZW|\end{array}$$

But $|AX| = |AN|$ and $|AY| = |AL| = |AN| + |NL|$, therefore

$$|XY| = 2|AN| + |NL|$$

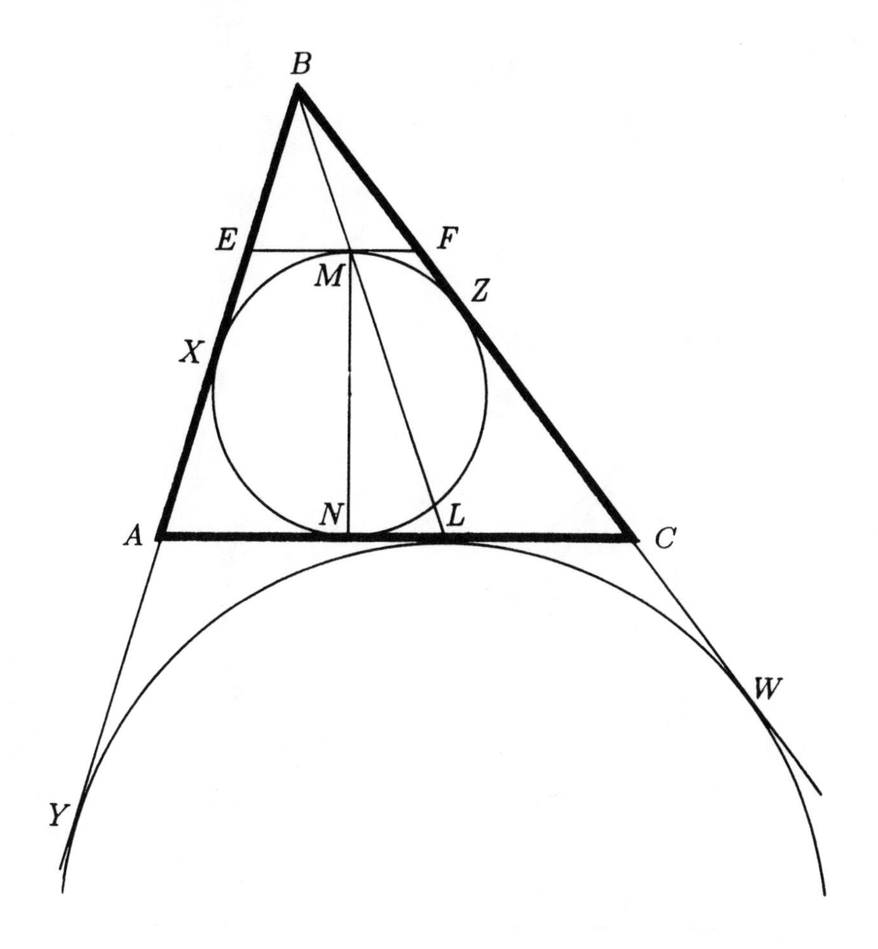

Figure 4.17

Similarly $|CZ| = |CN| = |NL| + |LC|$ and $|CW| = |CL|$, therefore,

$$|ZW| = 2|LC| + |NL|$$

Finally we have

$$2|AN| + |NL| = 2|LC| + |NL|,$$

i.e.

$$|AN| = |LC|$$

■

Second Solution

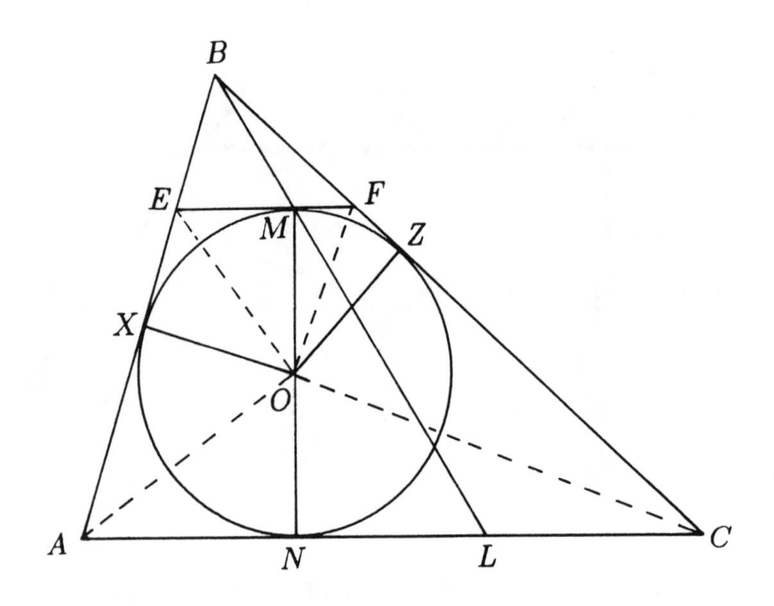

Figure 4.18

Let EF be the tangent line through M, then $EF \parallel AC$ (please see Figure 4.18). Let X and Z be the tangent points of the circle and BA and BC respectively. We connect A, C, F and E with the center O of the circle.

Let $m(\alpha)$ denote the measure of an angle α. Since $m(\angle OAE) + m(\angle OEA) = \frac{1}{2}m(\angle CAE) + \frac{1}{2}m(\angle FEA) = \frac{1}{2}\pi$, the triangle AOE is a right triangle; similarly the triangle FOC is a right triangle; therefore,

$$\begin{cases} r^2 = |OX|^2 = |EX| \cdot |XA| = |EM| \cdot |AN| \\ r^2 = |OZ|^2 = |FZ| \cdot |ZC| = |FM| \cdot |CN|, \end{cases}$$

thus

$$|EM| \cdot |AN| = |FM| \cdot |CN|,$$

i.e.

$$\frac{|EM|}{|MF|} = \frac{|CN|}{|AN|}$$

On the other hand, due to the homothetic transformation with the center B mapping M into L, we have

$$\frac{|EM|}{|MF|} = \frac{|AL|}{|LC|}$$

Combining the two equalities above, we get

$$\frac{|CN|}{|AN|} = \frac{|AL|}{|LC|}$$

If we denote $|AN| = a$, $|NL| = b$, $|LC| = c$, then the previous equality can be written as follows:

$$\frac{b+c}{a} = \frac{a+b}{c}$$

i.e.

$$a^2 + ab - c^2 - bc = 0,$$

or

$$(a-c)(a+b+c) = 0,$$

therefore,

$$a = c.$$

Thus $|AN| = |LC|$.

■

Problems

4.3.3 Prove that in any triangle, a bisector lies between the altitude and the median drawn from the same vertex of the triangle.

4.3.4 Prove that diagonals d_1, d_2 and sides a_1, a_2, a_3, a_4 of a parallelogram satisfy the following equality:

$$|d_1|^2 + |d_2|^2 = |a_1|^2 + |a_2|^2 + |a_3|^2 + |a_4|^2$$

4.3.5 An equilateral triangle ABC is inscribed in a circle. Prove that no matter where on the arc AB we chose a point P, $|PC| = |PA| + |PB|$.

4.3.6 Prove that the medians of any triangle are themselves the sides of a triangle.

4.3.7 (Ceva Theorem) The vertices A, B, C of a triangle ABC are connected by straight lines with the points M, N, K on opposite sides of the triangle (see Figure 4.19). If AM, BN and CK intersect in one point, then

$$\frac{|AK|}{|KB|} \cdot \frac{|BM|}{|MC|} \cdot \frac{|CN|}{|NA|} = 1,$$

and converse.

4.3.8 Prove that the straight line through the intersection of the extensions of the non-parallel sides of a given trapezoid and the intersection of its diagonals bisects both parallel bases of the trapezoid.

4.3.9 Prove that if the straight line through the intersection of the extensions of the opposite sides AB and CD of the given convex quadrilateral $ABCD$ and the intersection O of its diagonals AC and BD bisects its base AD or its base BC, then the given quadrilateral is a trapezoid, namely $BC \parallel AD$ (see Figure 4.20).

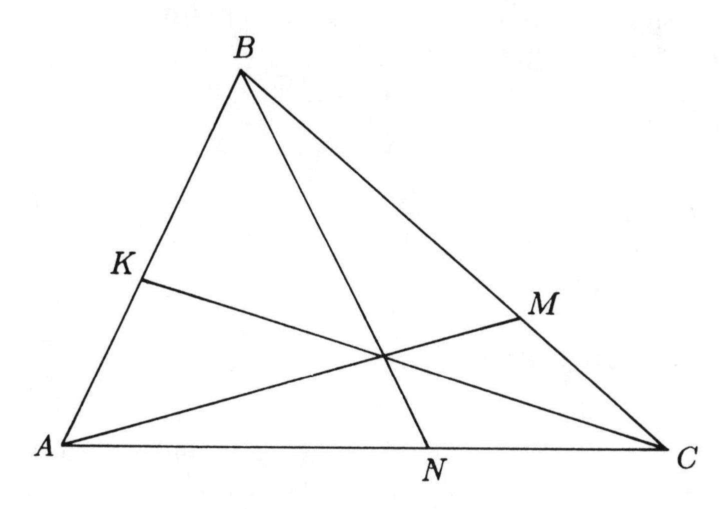

Figure 4.19

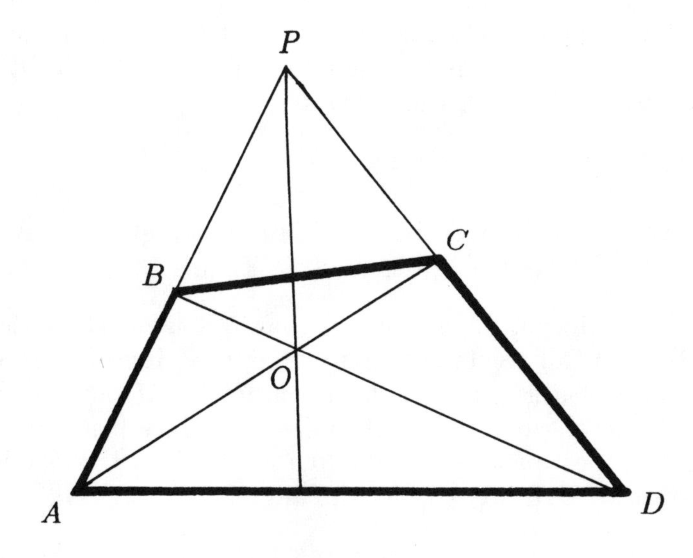

Figure 4.20

4.3.10 Prove that the altitudes of an arbitrary triangle intersect in one point.

4.3.11 (Euler straight line) Prove that the point M of intersection of medians, the center O of the circumscribed circle, and the point H of intersection of altitudes of a triangle lie on one straight line.

Moreover,

$$\frac{|\overrightarrow{OM}|}{|\overrightarrow{MH}|} = \frac{1}{2}$$

4.4 Constructions

In this section we will show how loci and transformations work on geometrical constructions. We will also look into constructions with limited means (compass, ruler).

4.4.1 Given an angle BAC, $0 < m(\angle BAC) < \pi$ and a point O inside it. Construct a straight line L, such that O is the midpoint of the segment cut out of the line L by the sides BA and BC of the given angle.

Analysis

Assume that MN is the required straight line, where M is on BA, and N is on BC (see Figure 4.21).

Let us forget for a moment that N lies on BC (Figure 4.22), then all we know about N is that it is the second endpoint of a segment with one endpoint M on the given line BA and with midpoint in O. Therefore (see Problem 4.1.5) N lies on the straight line S which is the result of rotation of BA about O by the angle π.

Now the intersection of S and BC will give us the point N.

Thus all we need to know is how to draw S. A straight line is determined by its two points, so if we pick two points

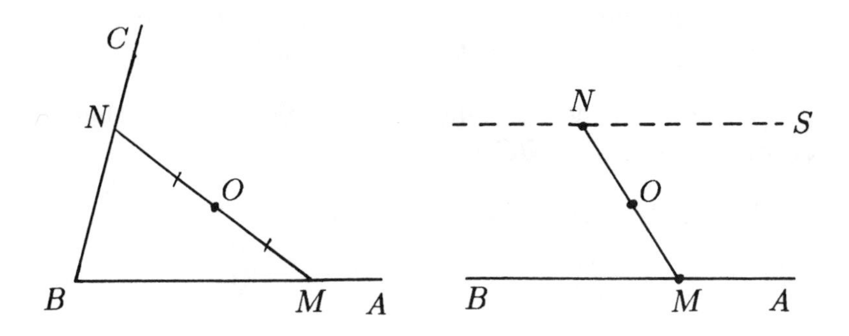

Figure 4.21 Figure 4.22

M_1, M_2 on the line BA and rotate each of them about O by π, we would get two points M_1', M_2' of the line S.

Construction

(1) Pick two distinct points M_1, M_2 on BA (see Figure 4.23).

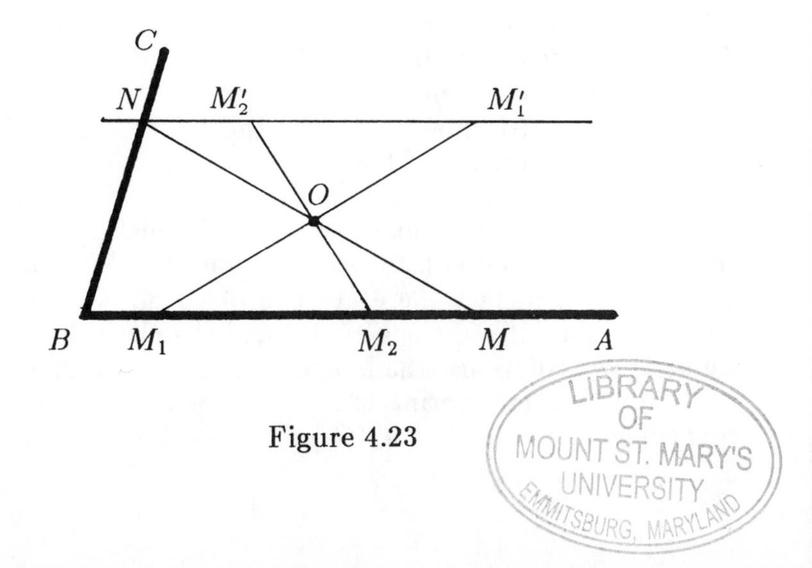

Figure 4.23

(2) Draw the straight line through M_1 and O, and mark a point M_1' on it, such that $|M_1O| = |OM_1'|$.

(3) Draw the straight line through M_2 and O, and mark a point M_2' on it, such that $|M_2O| = |OM_2'|$.

(4) Draw the straight line through M_1' and M_2'. Denote its intersection with BC by N.

(5) The required straight line is NO.

■

4.4.2 In the same setting (i.e., given an angle BAC, $0 < m(\angle BAC) < \pi$ and a point O inside it) construct a circle inscribed in the angle BAC and through the point O. How many solutions does the problem have?

Analysis

In order to construct the required circle we need to find its radius and locate the center. But the center will lie on the bisector AM of the angle BAC (Figure 4.24).

In fact the center of any circle inscribed in the angle BAC would lie on the bisector AM. Let us draw one such circle K, with the center in the point, say D. Then the homothetic transformation T with the center in A and an appropriate coefficient of homothety would map this circle into the required circle. But under the homothety T, the point O would be the image of a point, which is both on the circle K and also on the straight line AO. We can find such a point; in fact there are two of them, O_1 and O_2, which are the intersections of AO and the circle K. Thus we have two solutions: the image K_1 of the circle K under the homothetic transformation T_1 with the center A and the coefficient determined by $T_1(O_1) = O$; and the image K_2 of

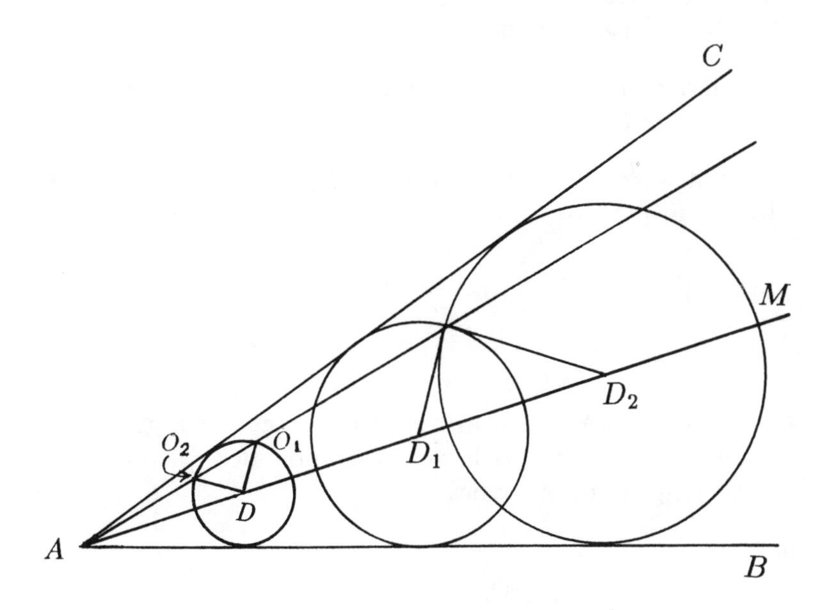

Figure 4.24

the circle K under the homothetic transformation T_2 with the center A and the coefficient determined by $T_2(O_2) = O$.

Construction Outline

We present here an outline of construction, and recommend the reader to describe completely the construction and provide all the proofs.

(1) Construct bisector AD of the angle BAC.

(2) Inscribe a circle K in the angle BAC; denote its center by D.

(3) Draw the straight line AO. Denote its intersections with the circle K by O_1 and O_2.

(4) Draw the straight line DO_1.

(5) Construct the straight line through O parallel to DO_1. Denote its intersection with AM by D_1.

(6) Draw the required circle K_1 with the center D_1 and the radius $|D_1O|$.

In order to receive the second solution K_2, just replace O_1 by O_2 in (4), (5) and (6).

■

4.4.3 Given segments $\bar{a}$ and $\bar{b}$. Construct the segment of the length $\sqrt[4]{|a|^4+|b|^4}$ with compass and ruler (please note: the unit length is not given!).

Analysis

$$\sqrt[4]{|a|^4+|b|^4} = \sqrt[4]{|a|^2\left(|a|^2+\frac{|b|^4}{|a|^2}\right)} = \sqrt{|a|\sqrt{|a|^2+\left(\frac{|b|^2}{|a|}\right)^2}}$$

This gives away the plan of our construction!

Construction

(1) Construct the segment of the length $|b|^2/|a|$. One way to do it is to construct a right triangle with the height $\bar{b}$ drawn from the right angle to the hypotenuse, and projection of one side on the hypotenuse equal to $\bar{a}$ (see Figure 4.25).

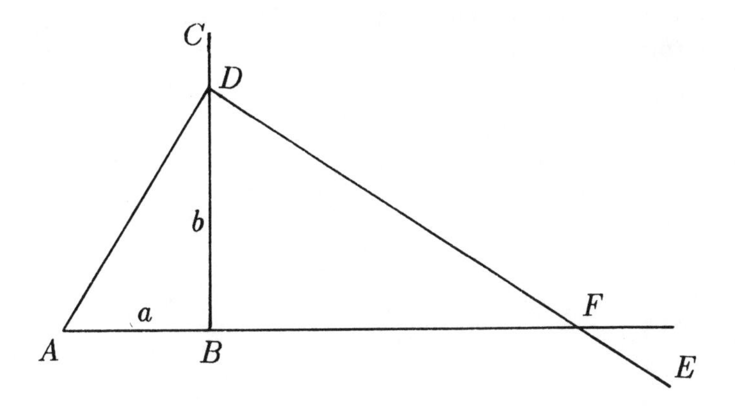

Figure 4.25

So, we draw:
(a) $\overline{AB}$, such that $|AB| = |a|$
(b) $CB \perp AB$
(c) plot D such that $|DB| = |b|$
(d) the straight line through A and D
(e) $DE \perp AD$
(f) denote the intersection of DE and AB by F

Then the length of $\overline{BF}$ is equal to $|b|^2/|a|$ (prove it!)

(2) Construct the segment $\bar{c}$ of the length

$$|c| = \sqrt{|a|^2 + \left(\frac{|b|^2}{|a|}\right)^2}$$

The hypotenuse of the right triangle with the legs of lengths $|a|$ (given) and $|b|^2/|a|$ (constructed above) has the required length.

(3) Construct the segment of the length $\sqrt{|a||c|}$ where $\bar{c}$ is constructed in (2). One way to do it is to construct a

right triangle with projections of legs on hypotenuse equal to $\overline{a}$ and $\overline{c}$, and take the altitude drawn from the right angle to the hypotenuse (see Figure 4.26):

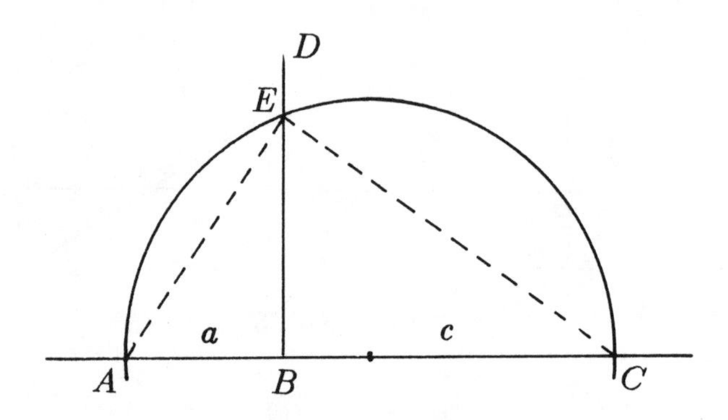

Figure 4.26

(a) Mark $|AB| = |a|$; $|BC| = |c|$
(b) Draw a circle on $\overline{AC}$ as on the diameter
(c) Draw $BD \perp AC$
(d) Denote an intersection of BD with the circle by E

Then the length of $\overline{BE}$ is equal to $\sqrt{|a||c|}$ (prove it!), and

$$\sqrt{|a||c|} = \sqrt{|a|\sqrt{|a|^2 + \left(\frac{|b|^2}{|a|}\right)^2}} = \sqrt{|a|^4 + |b|^4}$$

■

4.4.4 Given a circle K, its diameter $\overline{AB}$, and a point outside of K and not on the straight line AB. With a straight edge alone construct the perpendicular to AB through P. (Please note the center of the circle is not given!).

Analysis

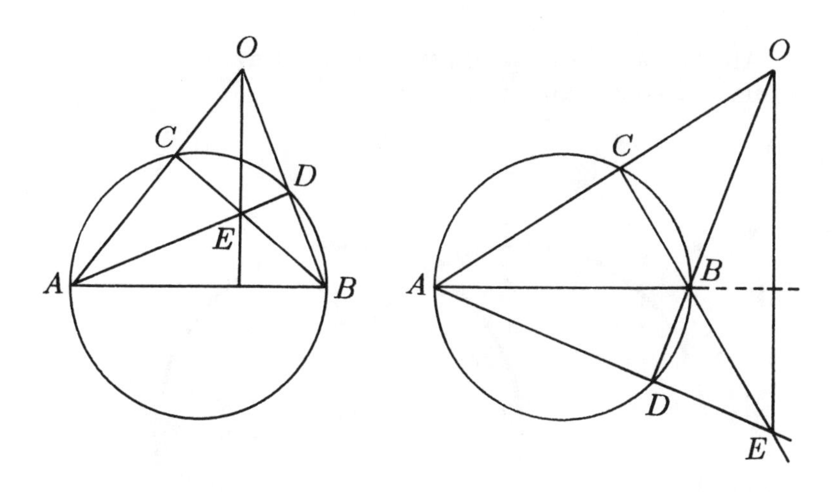

Figure 4.27 Figure 4.28

If we draw straight lines OA and OB and denote their intersections with the circle K by C and D respectively, then $m(\angle ACB) = m(\angle ADB) = \frac{\pi}{2}$ (please see both Figures 4.27 and 4.28 representing two possible cases: both $\angle OAB$ and $\angle OBA$ are acute, or one of them is obtuse. The third case when $\angle OAB$ or $\angle OBA$ is equal $\frac{\pi}{2}$ is trivial).

If we denote the intersection of AD and BC by E, the line OE will be the required perpendicular to AB because three altitudes of a triangle intersect in one point!

Construction is clear from the analysis and left for the reader.

■

4.4.5 With a compass alone double the given segment $\overline{AB}$.

First Solution

Analysis

All we can do to begin with is to draw two circles of radius $\overline{AB}$ with centers in A and B:

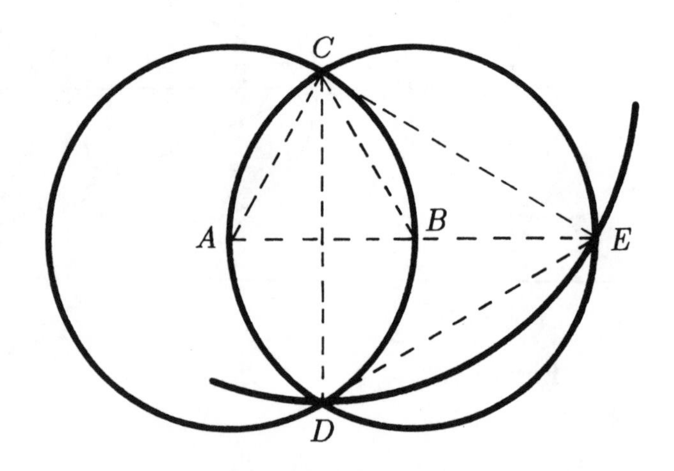

Figure 4.29

These two circles give us two new points C and D with $|CD| = |AB| \cdot \sqrt{3}$ (prove it!).

In order to find the required point E ($|AE| = 2|AB|$), we need to find two circles through E, which we can construct. But one we already have (please see Figure 4.29). The other is easy to notice due to equality $|CD| = |CE|$ (prove this too!).

Construction

(1) Draw the circles K_1 and K_2 of radius $|AB|$ with the centers in A and B respectively. Denote by C and D their intersections.

(2) Draw the circle K_3 of radius $|CD|$ with the center in C. Its intersection E with K_2 delivers to us the required segment $\overline{AE}$.

Proof is essentially done above by the reader (I hope).

■

Second solution is based on the observation that the side of a regular hexagon is equal to the radius of the circumscribed circle.

The implementation of this idea is left for the reader.

Problems

4.4.6 Given a straight line L and two points A, B on one side of it. Construct a circle through A and B tangent to L. How many solutions can the problem have?

4.4.7 Given segments $\bar{a}$ and $\bar{b}$, $|a| \geq |b|$. With compass and straight edge (and without the unit length) construct the segment of the length $\sqrt[4]{|a|^4 - |b|^4}$.

4.4.8 (Moscow Mathematical Olympiad, 1966) Divide the given segment $\bar{a}$ into six segments of equal length by drawing only 8 lines with compass and straight edge.

4.4.9 Given two parallel straight lines L_1 and L_2, and a point P. With a straight edge alone construct the line L parallel to L_1 and L_2 through P. (Hint: first solve Problem 4.3.9, then use it!)

4.4.10 With a compass alone divide the given segment in half. (Hint: use Problem 4.4.5 discussed in this section.)

4.5 Computations in Geometry

4.5.1 Given the lengths a, b, c of sides of a triangle T. Compute:

$$(h_a + h_b + h_c)\left(\frac{1}{h_a} + \frac{1}{h_b} + \frac{1}{h_c}\right),$$

where h_a, h_b, h_c are the lengths of the corresponding altitudes of T.

Solution

If S denotes the area of the triangle T, then $2S = ah_a = bh_b = ch_c$, and we get

$$(h_a + h_b + h_c)\left(\frac{1}{h_a} + \frac{1}{h_b} + \frac{1}{h_c}\right) =$$
$$\left(\frac{2S}{a} + \frac{2S}{b} + \frac{2S}{c}\right)\left(\frac{a}{2S} + \frac{b}{2S} + \frac{c}{2S}\right) =$$
$$(a+b+c)\left(\frac{1}{a} + \frac{1}{b} + \frac{1}{c}\right)$$

■

4.5.2 Let E, F, G be such points on the sides $\overline{AB}, \overline{BC}, \overline{CA}$ of the triangle ABC respectively, that

$$\frac{|AE|}{|EB|} = \frac{|BF|}{|FG|} = \frac{|CG|}{|GA|} = k,$$

where $0 < k < 1$. Find the ratio of the area of the triangle KML created by the intersections of the straight lines AF, BG, CE to the area of the triangle ABC (see Figure 4.30).

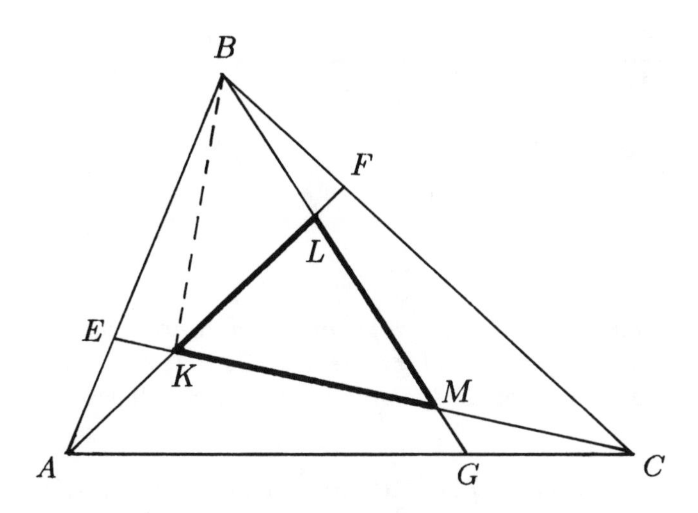

Figure 4.30

Solution

Denote the area of a triangle XYZ by $S(XYZ)$. Then by using several times the fact that the ratio of the areas of two triangles with the same heights (or the same bases) is equal to the ratio of their bases (the ratio of their heights, respectively), and $S(AEC) = S(AEK) + S(ACK)$, we get:

$$S(ACK) = \tfrac{1}{k}S(ABK) = \tfrac{k+1}{k^2}S(AEK) =$$

$$\frac{k+1}{k^2+k+1}S(AEC) = \frac{k}{k^2+k+1}S(ABC)$$

Similarly,

$$S(BLA) = S(CMB) = \frac{k}{k^2+k+1}S(ABC)$$

Therefore,

$$\frac{S(KLM)}{S(ABC)} = 1 - \frac{3k}{k^2+k+1} = \frac{(1-k)^2}{k^2+k+1} = \frac{(1-k)^3}{1-k^3}$$

■

Not only loci, transformations, and constructions help in solving computational problems. Sometimes computational problems provide the only clue for other types of problems.

4.5.3 (A. Soifer, 1971) Partition an arbitrary triangle by six straight cuts into such parts, from which it is possible to put together seven congruent triangles.

Analysis

In the previous problem we had

$$\frac{S(KLM)}{S(ABC)} = \frac{(1-k)^3}{1-k^3}$$

It is easy to see,that for $k = \frac{1}{2}$, $\frac{(1-k)^3}{1-k^3} = \frac{1}{7}$. So if we partition the sides of a triangle in the ratio 1:2,

$$\frac{|AE|}{|EB|} = \frac{|BF|}{|FC|} = \frac{|CG|}{|GA|} = \frac{1}{2}$$

and make cuts AF, BG, CE we will have one part, the triangle KLM of exactly the right size: $S(KLM) = \frac{1}{7}S(ABC)$.

We can also notice that if $k = \frac{1}{2}$, each of the three segments $\overline{AF}$, $\overline{BG}$ and $\overline{CE}$ is split by the two others in the ratio $3:3:1$ (counting from the angles of the triangle ABC).

Indeed the chain of the area equalities of the previous problem at $k = \frac{1}{2}$ gives us $S(AEK) = \frac{1}{7}S(AEC)$, therefore

$$|EK| = \tfrac{1}{7}|EC|. \qquad (*)$$

Similarly, $|GM| = \frac{1}{7}|GB|$, and $S(ABK) = S(ACM)$. But the same chain of equalities shows that $S(ABK) = \frac{1}{2}S(ACK)$, therefore $S(ACM) = \frac{1}{2}S(ACK)$, i.e.,

$$|CM| = \tfrac{1}{2}|CK|. \qquad (**)$$

The two equalities $(*)$ and $(**)$ above prove, that $|CM| : |MK| : |KE| = 3 : 3 : 1$.

Now it is not hard to find the remaining three cuts: we draw them through the points M, K, L parallel to AF, BG, CE respectively. They certainly complete the partition of the sides of the given triangle in the ratios 2:1:1:2.

Construction

See!

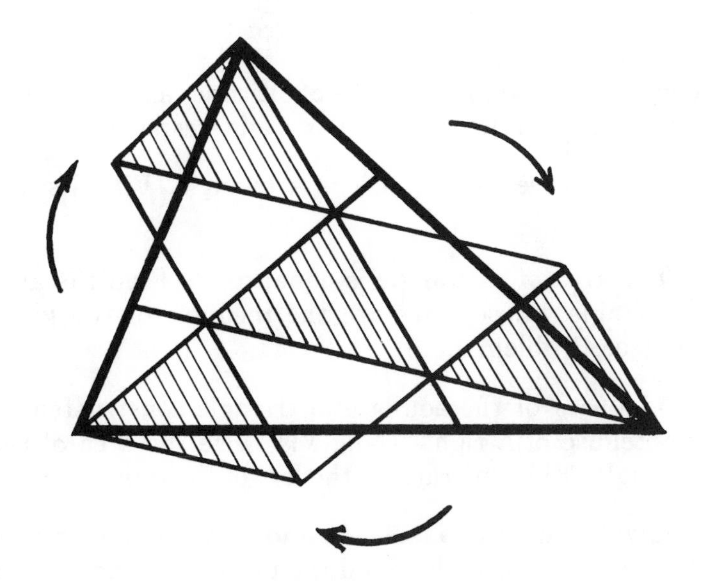

Figure 4.31

Proof is essentially contained in our detailed analysis.

■

Problems

4.5.4 Prove that,

$$h_a + h_b + h_c = \frac{ab + bc + ca}{2R},$$

where a, b, c, h_a, h_b, h_c, R are the lengths of sides, altitudes, and the radius of the superscribed circle of a given triangle, respectively.

4.5.5 Prove that,

$$\frac{1}{h_a} + \frac{1}{h_b} + \frac{1}{h_c} = \frac{1}{r}$$

where h_a, h_b, h_c, r are the lengths of altitudes and the radius of the inscribed circle of a given triangle, respectively.

4.5.6 Given three sides a, b, c of a triangle. Find its medians.

4.5.7 The area of a triangle is equal to S. Find the area of the triangle whose sides are the medians of the given triangle (see Problem 4.3.6).

4.5.8 The area of the equilateral triangle constructed on the hypotenuse of a right triangle is twice the area of the right triangle. Find the ratio of the legs of the right triangle.

4.5.9 Given the area S and the radius R of the circumscribed circle of a triangle. Find the product of the lengths of its three sides.

4.5.10 Given a right triangle ABC with the right angle A. The altitude $\overline{AK}$ is drawn to the hypotenuse; and from K the perpendiculars $\overline{KP}$ and $\overline{KT}$ are drawn to the legs $\overline{AB}$ and $\overline{AC}$ of the triangle ABC, respectively. If $|\overline{BP}| = m$, and $|CT| = n$, find the length of the hypotenuse $\overline{BC}$.

4.6 Maximum and Minimum in Geometry

4.6.1 Out of all triangles with the given base and area find the one of the minimal perimeter.

Analysis

The area S and the base $|BC| = a$ of a triangle ABC are given, but then we can figure out the length of the altitude h_a from the equation $S = \frac{1}{2}ah_a$.

But the locus of all points A, which are on the given distance h_a from the given straight line BC in the half-plane above BC is the straight line L parallel to BC on the distance h_a from BC (Problem 4.1.7).

Since the perimeter $P = |BA|+|AC|+|BC|$ and $|BC| = a$, the minimal perimeter is attained exactly when $|BA| + |AC|$ attains its minimum.

Now we can reformulate the problem as follows: for given points B, C and a straight line L find the point A on L which minimizes $|BA| + |AC|$ (please see Figure 4.32).

But we already solved this problem in Section 4.2 (Problem 4.2.1)!

In a particular case when $BC \parallel L$ (which we have here), the resulting triangle BAC is isosceles (prove it!).

Construction is left for the reader.

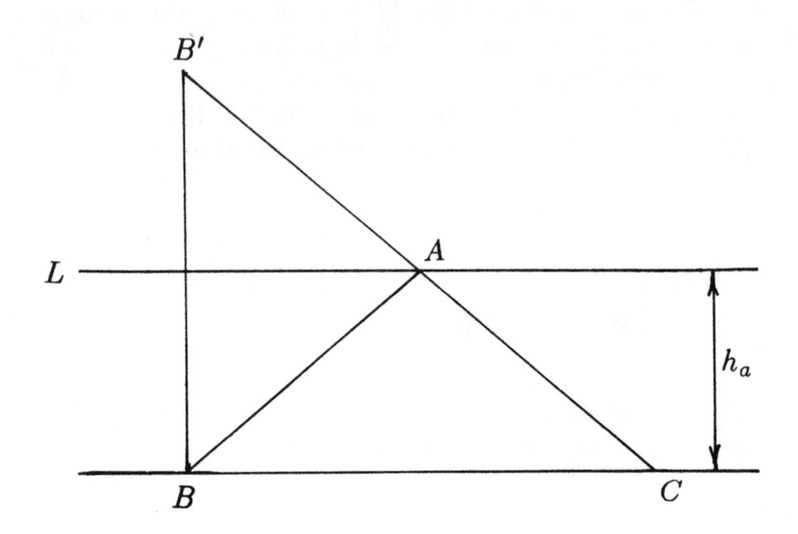

Figure 4.32

Proof is essentially contained in analysis above.

■

4.6.2 Given an angle ABC, $0 < m(\angle ABC) < \pi$ and a point O inside it. Find the straight line through O which cuts out of the angle ABC the triangle of the minimal area.

Analysis

The setting of this problem might remind you of Problem 4.4.1, which we discussed in Section 4.4. In that problem we constructed the straight line L, such that the point O was the midpoint of the segment $\overline{MN}$ cut out of L by the sides BA and BC of the given angle. Let us compare MN to any other straight line through O (see Figures 4.33 and 4.34).

Let $ND \parallel BC$ on Figure 4.33, and $ME \parallel BA$ on Fig-

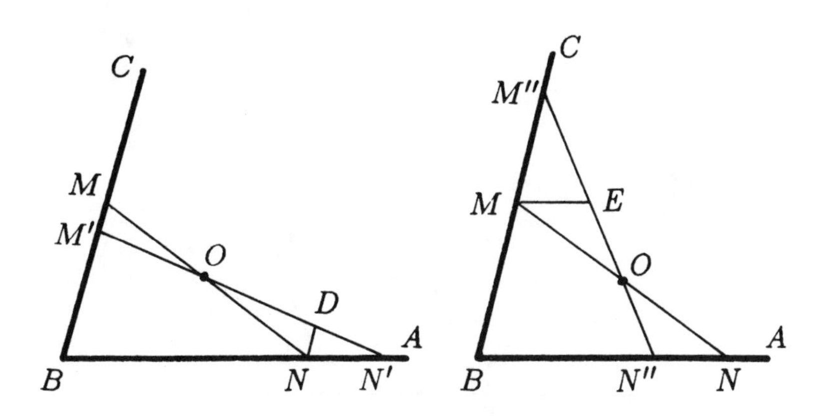

Figure 4.33 Figure 4.34

ure 4.34. Without any difficulties you can prove now (do!) that $S(MBN) < S(M'BN')$ and $S(MBN) < S(M''BN'')$, where $S(XYZ)$ denotes the area of a triangle XYZ.

Construction is already done in Problem 4.4.1!

Proof is contained in analysis.

■

4.6.3 In the setting of the previous problem, find the straight line through O, which cuts out of the angle ABC the triangle of the minimal perimeter.

Analysis

(a) For the time being let us forget about the point O and inscribe a circle K in the angle ABC. Denote the tangent points of K and BA and K and BC by M and N respectively (see Figure 4.35).

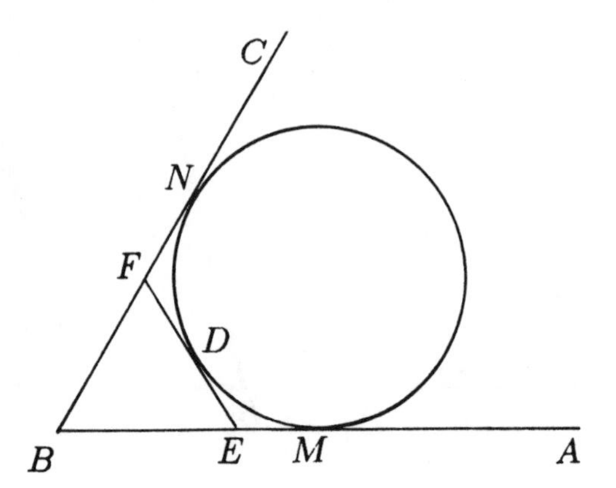

Figure 4.35

Let us now draw the tangent line EF through a point D of the circle K which is inside the triangle MBN. Since the lengths of the segments of two tangent lines drawn from a point to a circle are equal, we get: $|BN| = |BM|$; $|EM| = |ED|$; $|FD| = |FN|$.

Now we can figure out the perimeter $P(BEF)$ of the triangle BEF in terms of the length of the segment $\overline{BM}$ of the tangent line:

$$\begin{aligned} &P(BEF) = \\ &\quad (|BF| + |FD|) + (|BE| + |ED|) = \\ &\quad (|BF| + |FN|) + (|BE| + |EM|) = \\ &\quad |BN| + |BM| = 2|BM| \end{aligned}$$

(b) Now let us put back the point O, draw a straight line through it intersecting the sides BA and BC of the given angle in the points E and F respectively, and inscribe the circle K in the angle ABC, tangent to EF outside of the triangle EBF (see Figure 4.36).

From part (a) we know that $P(EBF) = 2|BM|$. Therefore, in order to minimize $P(EBF)$ we need to "push" the

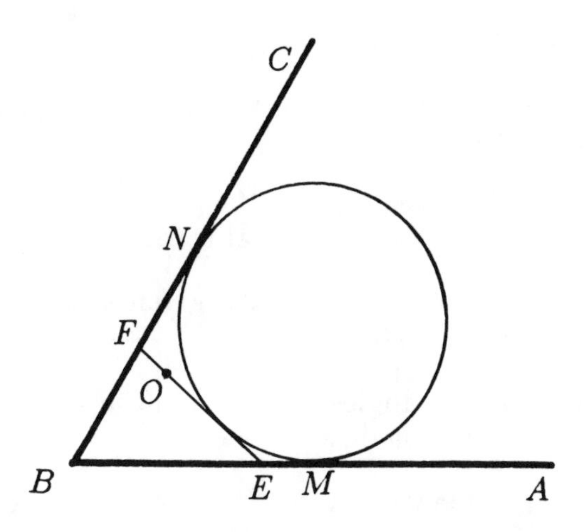

Figure 4.36

circle K in the angle ABC as much as possible, and thus minimize the length of the segment $\overline{BM}$ of the tangent line.

But we have a natural limit to our pushing: when the point O gets on the circle K we must stop!

Now the construction is clear: we construct a circle K through the given point O inscribed in the given angle ABC (Problem 4.4.2!), and draw the tangent line to K at the point O.

It is important to choose out of the two circles constructed in Problem 4.4.2 the larger one.

Construction is already done in the Problem 4.4.2!

Proof is essentially contained in our analysis above.

■

Problems

4.6.4 Out of all triangles with two given sides find the one of the maximal area.

4.6.5 Out of all triangles with the given side and opposite angle find the one of the maximal area.

4.6.6 Out of all rectangles of the given area find the one of the minimal perimeter.

4.6.7 Out of all rectangles inscribed in the given circle find the one of the maximal area.

4.6.8 Given three points A, B, P on the plane. Find the straight line L through P which maximizes the sum of distances from A and B to L.

4.6.9 Given the points A, B, P on the plane. Find the straight line through P which minimizes the sum of distances from A and B to L.

4.6.10 Out of all triangles inscribed in the given circle find the one of the maximal area.

4.6.11 Given a natural number n and a circle. Out of all inscribed convex n-gons find the one of the maximal area.

4.6.12 Out of all triangles inscribed in the given square find the one of the maximal area.

Chapter 5
Combinatorial Problems

5.1 Combinatorics of Existence

Clearly all the problems we discussed in Section 1.4 Pigeonhole Principle, delivered existence results, and thus belong here too.

There are, however, many other striking ideas used in combinatorial problems on existence. We will introduce here a few of them.

5.1.1 (Second Annual Colorado Springs Mathematical Olympiad, 1985) Each of the 49 entries of a square 7×7 table is filled by an integer between 1 and 7, so that each column contains all of the integers 1, 2, 3, 4, 5, 6, 7, and the table is symmetric with respect to its diagonal D going from its upper left corner to its lower right corner. Prove that this diagonal D has all of the integers 1, 2, 3, 4, 5, 6, 7 on it.

Solution

The total number of entries 1 in the table is odd (one per column). For every entry 1 off the diagonal D, the table contains 1 in the square symmetric to the first one with respect to D. Therefore, the number of entries 1 off the diagonal D in the table is even, thus D contains at least one entry 1.

Similarly D contains at least one 2, at least one 3, ..., at least one 7.

■

5.1.2 (Ramsey) Prove that in any party of six people, there are three mutual acquaintances or three mutual non-acquaintances.

Solution

(a) It is convenient to record the information about acquaintances of a group of six people by a combination of two diagrams D_1 and D_2. In both diagrams we represent every person by a vertex of a regular hexagon, and two vertices of D_1 are connected by a segment if and only if they correspond to two people who are acquainted.

Two vertices of D_2 are connected by a segment if and only if they correspond to two people who are not acquainted.

Please note that any two vertices are connected in exactly one of the two diagrams D_1, D_2.

The given problem is now equivalent to proving that at least one of the two diagrams D_1, D_2 contains a triangle!

(b) Let us fix the same vertex A in both diagrams D_1 and D_2. A is connected with each of the five other vertices either in D_1 or in D_2, therefore, it must be connected with at least three of the five vertices in D_1 or in D_2 (Pigeonhole Principle with five pigeons and two holes!).

Due to the symmetry of the problem, we can assume without loss of generality that A is connected with the ver-

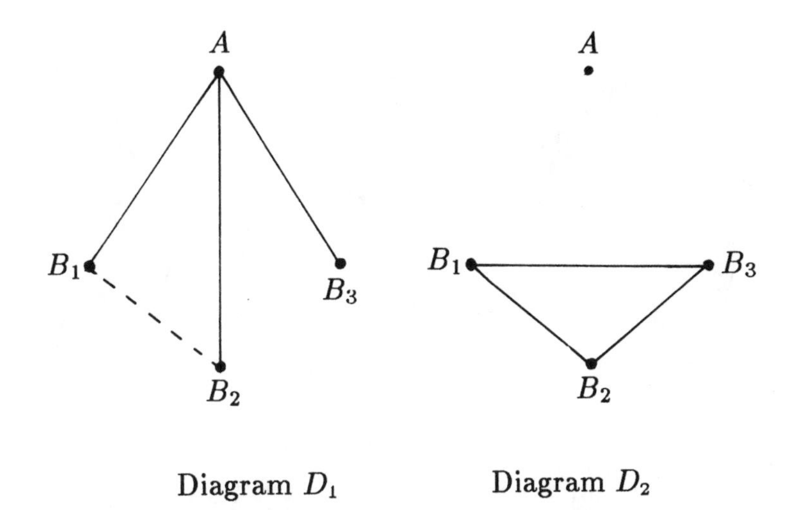

Diagram D_1 Diagram D_2

tices B_1, B_2 and B_3 in D_1. If any two of the vertices B_1, B_2, B_3 are connected in D_1, then these two vertices and A are the vertices of a triangle contained in D_1.

If no two of the vertices B_1, B_2, B_3 are connected in D_1, then the triangle B_1, B_2, B_3 is contained in the diagram D_2.

■

It is interesting to notice that six is the smallest number of people in the company to guarantee the above result. A party of five people might have neither three mutual acquaintances or three mutual non-acquaintances. This can be proven by the following two diagrams neither of which contains a triangle (please see Diagram of Acquaintances and Diagram of Non-Acquaintances).

Once again I would like to discuss the same problem (please see Problem 1.4.5 in Section 1.4).

5.1.3 Forty-one rooks are placed on a 10×10 chessboard.

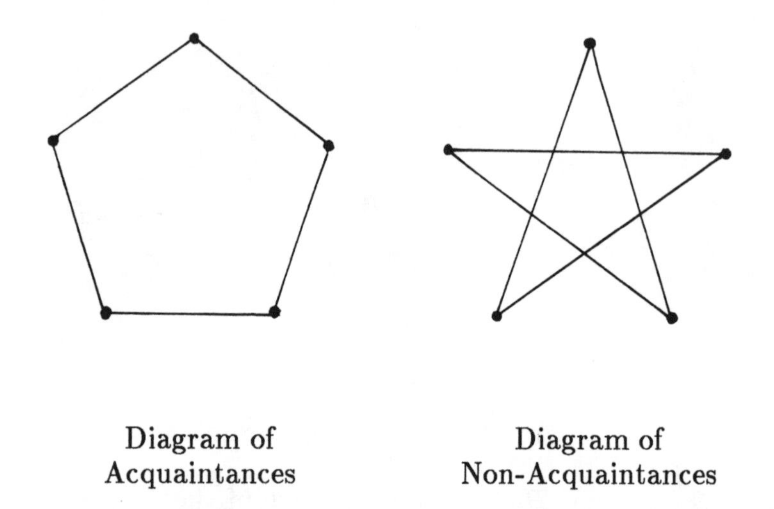

Diagram of Acquaintances

Diagram of Non-Acquaintances

Prove that you can choose five of them which are not attacking each other.

Second Solution

This solution was found by George Berry in 1984. Suppose that at most you could choose four rooks that are not attacking each other. Select any four rooks that do not attack each other. We will call these four the key rooks.

Now consider an operation on the chessboard that exchanges two rows or two columns of the board. This exchange does not affect the "attacks" of any of the rooks.

We can use this to "normalize" the chessboard. That is, to put the four key rooks into locations (0,0), (1,1), (2,2) and (3,3), where the first coordinate determines the row, and the second coordinate the column of a rook.

The normalization procedure can be done as follows: locate any one key rook and exchange its row with row 0 and its column with column 0; do likewise for the other three key rooks (moving them into (1,1), etc.). At the end of normalization, the chessboard looks like this:

	0	1	2	3	4	5	6	7	8	9
0	K	?	?	?	?	?	?	?	?	?
1	?	K	?	?	?	?	?	?	?	?
2	?	?	K	?	?	?	?	?	?	?
3	?	?	?	K	?	?	?	?	?	?
4	?	?	?	?	-	-	-	-	-	-
5	?	?	?	?	-	-	-	-	-	-
6	?	?	?	?	-	-	-	-	-	-
7	?	?	?	?	-	-	-	-	-	-
8	?	?	?	?	-	-	-	-	-	-
9	?	?	?	?	-	-	-	-	-	-

- = there is no rook
K = occupied by key rook
? = don't know whether there is a rook there or not

Consider the square of the chessboard with coordinates (n, m) or (m, n), such that $0 \leq n \leq 3$ and $3 \leq m \leq 9$. There are 48 such squares. We will call these the outside squares of the board.

Select a pair of outside squares with coordinates (n, m) and (m, n). Is it possible for both of these squares to contain a rook? No, for if both squares had rooks in them then we could select a new set of key rooks by taking the rooks on those two squares along with three of the four key rooks [omitting the one at (n, n)]. This would form a new set of *five* key rooks: we eliminated the only one of the the original key rooks that attacked (n, m) and (m, n); and obviously (n, m) does not attack (m, n). However a set of *five* key rooks is impossible by the original assumption that the largest possible set of key rooks was four.

Therefore, only one of the pair of outside squares (n, m) and (m, n) is occupied. That means that at most there are 24 rooks on the outside squares (half the squares are occupied). Where added to the 16 rooks (at most on the "inside" squares), we have a total of 40 rooks accounted for.

That is one less than the 41 rooks given in the problem, so the assumption that there are at most four key rooks must be false.

■

Problems

5.1.4 Is there a polyhedron with an odd number of faces, and each face with an odd number of sides?

5.1.5 Prove that the maximal number of bishops, which can be placed on the $n \times n$ board without attacking each other is equal to $2n - 2$.

(Two bishops attack each other if they are on the same diagonal of the board.)

5.1.6 Find the minimal number of rooks which have to be placed on a chessboard, so that every square is attacked by at least two rooks. (A rook attacks the square it is on, and all the squares in the same row and column, including the ones blocked or occupied by other rooks.)

5.1.7 A cycle of acquaintances is a group of n people, such that the first and the second persons are acquainted, the second and the third persons are acquainted, etc., the nth and the first persons are acquainted.

Prove that any party of people in which any cycle of acquaintances consists of an even number of people, can be partitioned into two groups, such that either group consists of mutual non-acquaintances.

5.1.8 Prove the converse of the statement of Problem 5.1.7.

5.2 How Can Coloring Solve Mathematical Problems?

5.2.1 A bureaucratic institution has one entrance, one exit, and a door in the middle of every interior wall of every room (see the plan on Figure 5.1.)

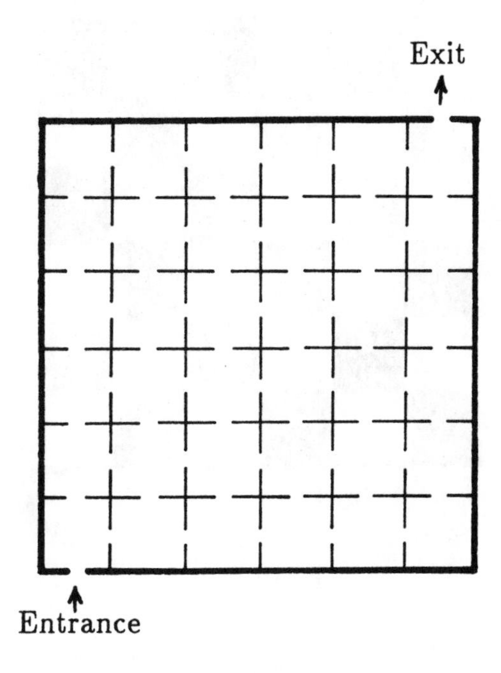

Figure 5.1. Plan of the Institution

In order to receive a certificate, one has to enter the building, visit exactly once every room, and exit the building.

Is there a way to receive a certificate?

Solution

Let us color (!) the plan in two colors in a chessboard fashion (see Figure 5.2).

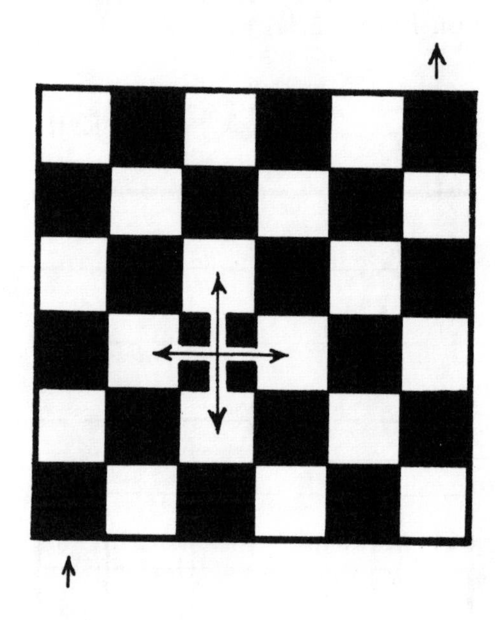

Figure 5.2

This coloring has a nice property: from a black room one can only go into a white room, and from a white room one can only enter a black room.

While walking through the building we will therefore alternate colors of rooms:

$$\text{black} \Rightarrow \text{white} \Rightarrow \text{black} \Rightarrow \text{white} \Rightarrow \ldots$$

Assume now that there is a required walk coming through every room once, beginning at the entry corner and ending at the exit corner. Since this walk begins at a black room, and alternates colors of the rooms on the way, it must end up on a white room (we have equal numbers of black and white rooms!), but the exit room is black. We arrived at contradiction, showing that one can not get a certificate out of this institution.

What else could you expect from bureaucracy!

■

We say that a figure $\mathcal{F}$ is tiled if it is completely covered by tiles without tiles overlapping or sticking out of the border of $\mathcal{F}$.

5.2.2 Find all pairs (m, n) of natural numbers such that the $m \times n$ checkerboard can be tiled by linear k-mino? (i.e., rectangles of the size $1 \times k$).

First solution

If m or n is a multiple of k then the $m \times n$ board can be tiled by linear k-mino (show!).

Assume now that neither of m or n is a multiple of k, but the board however can be tiled by linear k-mino. Let us color the board diagonally in k colors with cyclic permutation of colors (see Figures 5.3 and 5.4).

This coloring (diagonal cyclic k-coloring) has a remarkable property: no matter how a linear k-mino is placed on the board, horizontally or vertically, it will cover exactly one square of each of the k colors.

Since by assumption the board can be tiled by linear k-mino, and every k-mino covers an equal number of squares of every color (i.e. one square of each color), the board contains an equal number of squares of each of the k colors. It is not difficult to prove (do!), that for any natural numbers m and k there are non-negative integers q and r_1, such that

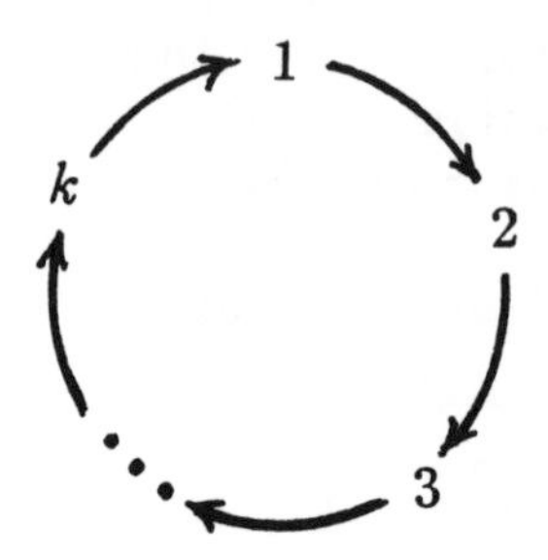

Cyclic permutation of k colors
Figure 5.3

Diagonal cyclic k-coloring for $k = 3$
Figure 5.4

$0 \leq r_1 \leq \frac{k}{2}$ and

$$m = kq + r_1$$

or

$$m = kq - r_1$$

Accordingly, we will consider two cases.

1. $m = kq+r_1$. We cut the given board into two rectangular boards: $kq \times n$ and $r_1 \times n$. Since the $nq \times n$ board can be tiled by linear k-mino, it contains an equal number of squares of each color. The given board too contains an equal number of squares of every color, therefore the $r_1 \times n$ board has the same property.

2. $m = kq - r_1$. In this case we attach the $r_1 \times n$ board to the given board to obtain the $kq \times n$ checker board, and extend the coloring of the given board to a diagonal cyclic k-coloring of the $kq \times n$ board. The $kq \times n$ board can be tiled by linear k-mino, therefore it contains an equal number of squares of every color. The given $m \times n$ board contains an equal number of squares of every color, therefore the $r_1 \times n$ board contains an equal number of squares of every color.

Thus in both cases we got the $r_1 \times n$ board with diagonal cyclic k-coloring, which contains an equal number of squares of every color. Let us turn this board now by a 90° angle, i.e. consider the $n \times r_1$ board, and apply to it all of the above reasoning. As a result we will get the $r_2 \times r_1$ board, where $0 < r_2 \leq \frac{k}{2}$, which contains an equal number of squares of every color.

On the other hand, the number of color diagonals in the $r_2 \times r_1$ board is equal to $r_2 + r_1 - 1$, and

$$r_2 + r_1 - 1 \leq \frac{k}{2} + \frac{k}{2} - 1 < k,$$

therefore at least one of the k colors is not present at all in the $k_2 \times k_1$ board!

Thus we proved that $m \times n$ board can be tiled by linear k-mino if and only if m or n is a multiple of k.

■

The following notation will be helpful for us in the second solution of this problem: *instead of writing "the remainders upon dividing numbers $a_1, a_2, \ldots, a_n$ by n are equal" we will simply write:*

$$a_1 \equiv a_2 \equiv \cdots \equiv a_n \pmod{n}$$

It is easy to prove (do!) that $a_1 \equiv a_2 \pmod{n}$ if and only if $a_1 - a_2$ is a multiple of n.

Second Solution

Assume that neither of m, n is a multiple of k, i.e.

$$m = kq_1 + r_1; \quad 0 < r_1 < k$$
$$n = kq_2 + r_2; \quad 0 < r_2 < k,$$

but the board is tiled by linear k-mino.

Let us color the columns of the board each in one of the k colors with cyclic permutation of colors (see Figures 5.3 and 5.5), and denote by $S_1, S_2, \ldots, S_k$ numbers of squares of the board colored in the 1st, 2nd, ..., k-th colors respectively.

This coloring (column cyclic k-coloring) has an interesting property: if a linear k-mino is placed on the board vertically, it would cover k squares of the same color; if it is placed on the board horizontally, it would cover exactly one square of every color. By assumption the board is tiled by linear k-mino, therefore

$$S_1 \equiv S_2 \equiv \cdots \equiv S_k \pmod{k}.$$

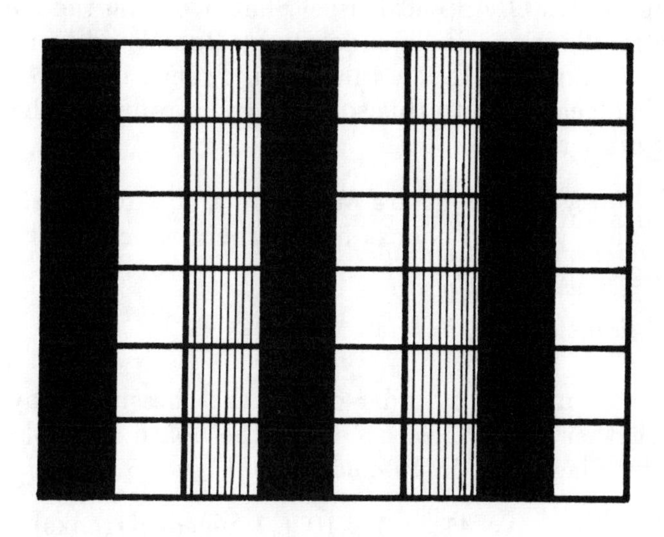

Column cyclic k-coloring for $k = 3$
Figure 5.5

On the other hand, you can notice that we have one column more of the r_2th color than of the $(r_2 + 1)$st color, i.e.

$$S_{r_2} - S_{r_2+1} = m,$$

therefore,

$$S_{r_2} \not\equiv S_{r_2+1} (\text{mod } k).$$

The contradiction we obtained proves that divisibility of m or n by k is a necessary condition for the $m \times n$ board to be "tileable" by linear k-mino. It is also sufficient.

■

For the third time I would like to discuss the same problem (please see Problem 1.4.5 in Section 1.4 and Problem

5.1.3 in Section 5.1), and present the third solution of it, obtained by the winner of the First Annual Colorado Springs Mathematical Olympiad Russel Shaffer during the competition in 1984; and independently by Bob Wood from Colorado Springs and Luc Miller from France in 1986. Bob added the elegance to this solution by introducing the idea of a cylinder.

5.2.3 Forty-one rooks are placed on a 10×10 chessboard. Prove that you can choose five of them which are not attacking each other.

Third Solution

Let us make a cylinder out of the chessboard by gluing together two opposite sides of the board and color the cylinder diagonally in 10 colors (see Figure 5.6)

Now we have $41 = 4 \times 10 + 1$ pigeons (rooks) in 10 pigeonholes (one-color diagonals), therefore there is at least one hole containing at least 5 pigeons. But the 5 rooks located on the same one-color diagonal do not attack each other!

■

Problems

5.2.4 Given an $m \times n$ rectangular chessboard and dominoes (linear k-minos with $k = 2$). A pair of distinct squares of the board is called *good*, if the figure resulting from the cutting of these two squares out of the board can be tiled by dominoes. Find all the *good* pairs.

5.2.5 Find all natural numbers n, such that a square $n \times n$ chessboard can be tiled by T-tetramino (please see Figure 5.7).

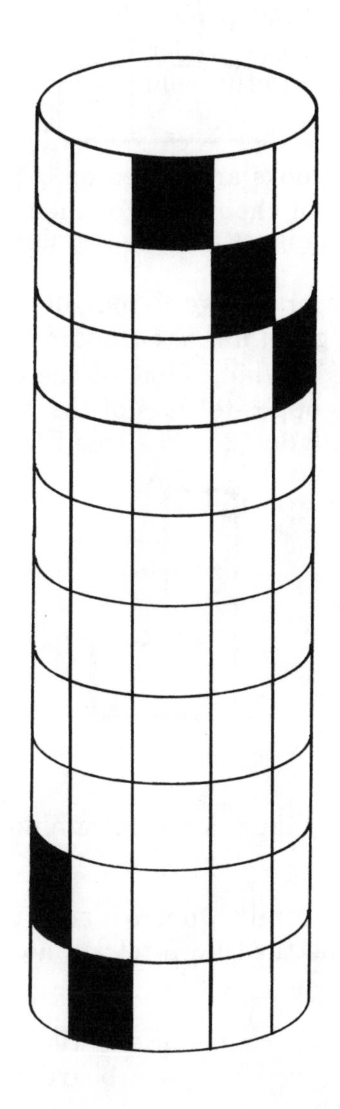

One out of the ten one-color diagonals is shown in black
Figure 5.6

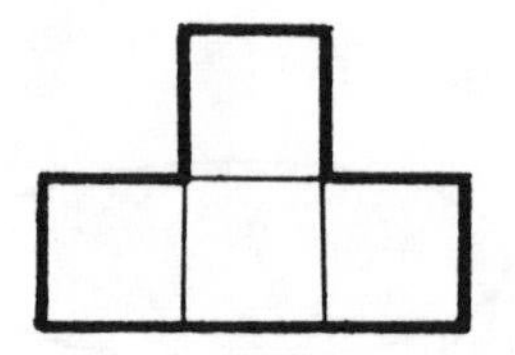

Figure 5.7 T-tetramino

Will the answer change if you tile a cylindrical board made out of the given flat $n \times n$ board?

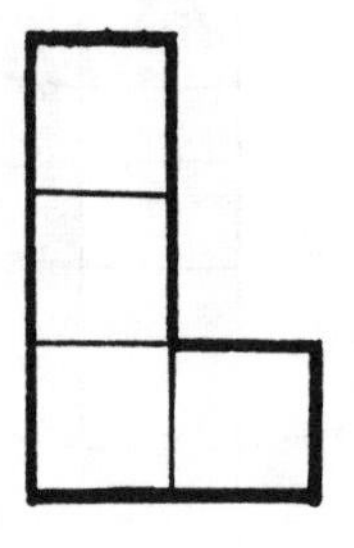

Figure 5.8 L-tetramino

5.2.6 Find all natural numbers n such that a square $n \times n$ chessboard can be tiled by L-tetramino (please see Figure 5.8).

Will the answer change if you tile a cylindrical board made out of the given flat $n \times n$ board?

5.3 Combinatorics of Sets

We will use standard notations of sets:

$a \in A$ - denotes "a is an element of the set A"
$a \notin A$ - denotes "a is not an element of the set A"
$A \cup B$ - union of sets A and B
$A \cap B$ - intersection of sets A and B
$A \backslash B$ - denotes the set of all elements x, such that $x \in A$ and $x \notin B$
$|A|$ - denotes the number of elements of a finite set A.

Let n and k be non-negative integers, and $k \leq n$. The symbol $\binom{n}{k}$ denotes the number of k-element subsets of an n-element set.

5.3.1 Prove that for any non-negative integers n and k, $k \leq n$, the following equality holds:

$$\binom{n+1}{k+1} = \binom{n}{k} + \binom{n}{k+1}$$

Solution

Let us fix one element a of the given $(n+1)$-element set S. We can partition all $(k+1)$-element subsets into those which contain a, and those which do not.

But if a is one element of a subset L which is to consist of $k+1$ elements, we need to select k more elements out of $S \setminus \{a\}$, so the number of such subsets L is by definition $\binom{n}{k}$.

If a is not an element of a subset M which is to consist of $k+1$ elements, then we need to select all $k+1$ elements out of the set $S \setminus \{a\}$, so the number of such subsets M is by definition $\binom{n}{k+1}$.

So the total number of $(k+1)$–element subsets of the given $(n+1)$–element set S is $\binom{n}{k} + \binom{n}{k+1}$.

On the other hand, by definition it is also equal to $\binom{n+1}{k+1}$.

■

5.3.2 Prove that the following equality holds for any positive integer n:

$$\binom{n}{0} + \binom{n}{1} + \binom{n}{2} + \cdots + \binom{n}{n} = 2^n$$

Solution

(a) Let S be an n–element set. Since $\binom{n}{0}$ denotes the number of 0–element subsets of S, $\binom{n}{1}$ denotes the number of 1–element subsets of S, etc., the sum

$$\binom{n}{0} + \binom{n}{1} + \binom{n}{2} + \cdots + \binom{n}{n}$$

is equal to the total number $|P(S)|$ of subsets of the set S.

(b) All there is left to prove is that the number $|P(S)|$ of subsets of the set S is equal to 2^n. One way to define a subset S_0 of the set S is to tell about each of n elements of S whether or not it is an element of S_0.

Elements	1st	2nd	$\cdots$	nth
Yes or No				

Figure 5.9

Thus we have two options (yes, no) for the 1st block, two options for the second block, ..., two options for the nth block of the Figure 5.9, i.e.

$$|P(S)| = \overbrace{2 \cdot 2 \cdot \cdots \cdot 2}^{n \text{ factors}} = 2^n$$

■

5.3.3 (MATHCOUNTS, National Competition, 1985). A collection of letters consists of n D's and r C's. Find the number P of different words (sequences) that can be formed from the D's and C's if each sequence must contain n D's (and not necessarily all C's).

Solution

I would like to share with you a solution I found in May, 1985. First let us add one more letter D, and find out the number of words that can be composed of exactly $n+1$ letters D and r letters C.

These words have $r+n+1$ letters:

1st	2nd	$\cdots$	$(r+n+1)$st

In order to uniquely determine such a word we have to pick the positions of letters D and fill the rest with C's, i.e. we need to pick an $(n+1)$–element subset out of the $(r+n+1)$–element set $\{1,2,\ldots,r+n+1\}$. Therefore the number R of such words is equal:

$$R = \binom{r+n+1}{n+1}$$

All there is left to do is to notice that the required number P is equal to R.

Indeed let us take a word W consisting of $n+1$ letters D and r letters C.

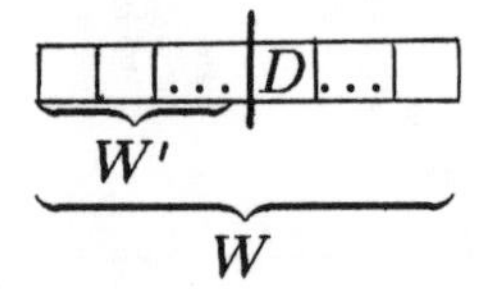

We locate the first D from the right and cut out the end of the word beginning with this D. The remaining part W' of the word W certainly contains n letters D and no more than r letters C. It is easy to show (do!) that two distinct words W_1, W_2 produce distinct remaining parts W_1', W_2' thus completing the proof that

$$P = R = \binom{r+n+1}{n+1}$$

So, adding an extra D and later cutting it off proves to be productive!

■

Problems

5.3.4 Prove that the following equality holds for any non-negative integers n and k, $k+1<n$:

$$\binom{n+2}{k+2} = \binom{n}{k} + 2\binom{n}{k+1} + \binom{n}{k+2}$$

5.3.5 How many ways are there to deliver 7 letters if there are 3 couriers and any letter can be given to any courier?

5.3.6 Given n points on a circle, how many quadrangles (not necessarily convex) can be inscribed in the circle with the vertices in the given points? How many of them are convex?

5.3.7 Find the number of intersections of diagonals of a convex n-gon P inside P, if no two diagonals are parallel and no three intersect in one point. Find the number of intersections of diagonals of P outside of P.

5.4 A Problem on Combinatorial Geometry

I would like to conclude these notes by offering you a problem. In 1970 when I raised and solved this problem, its main part was originally selected by the judges for the 9th grade competition of the Soviet Union National Mathematical Olympiad, but Academician A. N. Kolmogorov did not approve it as being too difficult for a four hour five problem competition. Of course he was right.

Find all natural numbers n, such that every triangle can be cut into n triangles similar to each other.

This and other problems of combinational geometry will be discussed in my next book which I am going to call **How to Cut a Triangle?**.

The main goal of this book will be to demonstrate synthesis, to show how ideas from various areas of mathematics such as geometry, algebra, trigonometry, linear algebra, extensions of rings, get together in the solution of a problem, to give a mini-model of mathematical research.

Literature

1. E.B. Dynkin, C.A. Molchanov, A. L. Tolpygo, A. K. Rozental. *Mathematicheskie Zadachi (Mathematical Problems)*, Nauka, Moscow, 1965, (Russian).

2. V. A. Krechmar. *Zadachnik no Algebre (Problem Book on Algebra)*, Nauka, Moscow, 1964, (Russian).

3. A. A. Leman. *Sbornik Zadach Moskovskich Matematicheskich Olimpiad. (Problem Book of Moscow Mathematical Olympiads)*, Prosvetshenie, 1965. (Russian).

4. Michel de Montaigne. *Essayes, John Florio's Translation*, The Modern Library, New York.

5. M. Popruzhenko. *Sbornik Geometricheskich Zadach (Selection of Geometrical Problems)*, OGIZ, Moscow, 1936, (Russian).

6. A. Soifer, *Problem M65*, Kvant #1 (1971), p. 39; and Kvant #10 (1971) pp. 36-37 (Russian).

7. A. Soifer, *Kletchatye Doski i Polimino (Checker Boards and Polymino)*, Kvant #11 (1972), pp. 2-10, (Russian).

8. A. Soifer and S. G. Slobodnik, *Problem M236*, Kvant #12 (1973), p. 29 (Russian).

9. N. Vasiliev and V. Gutemacher. *Zadachi po Geometrii (Problems on Geometry)*, Moscow University, 1965, (Russian).

10 N. Ja. Vilenkin. *Populiarnaya Kombinatorika (Popular Combinatorics)*, Nauka, 1975, (Russian).